AF576012

Recovery, Isolation and Characterization on Food Proteins

Recovery, Isolation and Characterization on Food Proteins

Editors

Ute Schweiggert-Weisz
Emanuele Zannini

MDPI • Basel • Beijing • Wuhan • Barcelona • Belgrade • Manchester • Tokyo • Cluj • Tianjin

Editors
Ute Schweiggert-Weisz
Fraunhofer Institute for Process Engineering and Packaging (IVV)
Germany

Emanuele Zannini
University College Cork
Ireland

Editorial Office
MDPI
St. Alban-Anlage 66
4052 Basel, Switzerland

This is a reprint of articles from the Special Issue published online in the open access journal *Foods* (ISSN 2304-8158) (available at: https://www.mdpi.com/journal/foods/special_issues/Recovery_Isolation_Characterization_Food_Proteins).

For citation purposes, cite each article independently as indicated on the article page online and as indicated below:

LastName, A.A.; LastName, B.B.; LastName, C.C. Article Title. *Journal Name* **Year**, *Volume Number*, Page Range.

ISBN 978-3-0365-4631-5 (Hbk)
ISBN 978-3-0365-4632-2 (PDF)

Contents

About the Editors

Ute Schweiggert-Weisz

Ute Schweiggert-Weisz studied Food Technology at the University of Hohenheim. During her PhD, which was also in Food Technology, she focused on the development of alternative processes for the production of spices. After graduation, she moved to Freising in 2003 to work at the Fraunhofer Institute for Process Engineering and Packaging. Active in various management positions, she worked intensively on the extraction, characterisation, modification, and application of plant proteins in various food products. In 2020, she joined the Institute of Nutritional and Food Sciences at the University of Bonn, where she heads the Food Sciences Working Group. In Bonn, she is currently expanding her research on plant proteins. She is particularly interested in protein crops, which have so far been of minor importance. In addition to analysing certain physicochemical properties, she would also like to dive deeper into the nutritional—physiological properties of plant proteins.

Emanuele Zannini

Emanuele Zannini studied Agricultural Science at the Universita' Politecnica delle Marche. During his PhD in Applied Biomolecular Science, he focused on the selection of antifungal lactic acid bacteria as biopreservatives in food products. In 2008/2009, he was Contract Professor of Instrumental Chemical Analysis at Faculty of Agriculture, Ancona. Since November 2009, he has been working as a Senior Researcher at the School of Food and Nutritional Sciences, UCC. His research projects focus on the functional characterisation of lactic acid bacteria for use as starter cultures or protective cultures in food and the production of hydrocolloids and antimicrobial compounds by lactic acid bacteria in cereal-based products. A new area of scientific interest is medical nutritional therapy. Various research activities on food microbiology, engineering, and nutrition are being conducted to formulate food products able to address consumers' preferences and dietary needs and complement medical therapy requirements.

MDPI

Editorial

Recovery, Isolation, and Characterization of Food Proteins

Ute Schweiggert-Weisz [1,2,*] and Emanuele Zannini [3]

1 Institute for Nutritional and Food Sciences, University of Bonn, 53113 Bonn, Germany
2 Department Food Process Development, Fraunhofer Institute for Process Engineering and Packaging (IVV), 85354 Freising, Germany
3 Cereal and Beverage Science Research Group, School of Food and Nutritional Sciences, University College Cork, College Road, T12K8AF Cork, Ireland; e.zannini@ucc.ie
* Correspondence: ute.weisz@ivv.fraunhofer.de

Citation: Schweiggert-Weisz, U.; Zannini, E. Recovery, Isolation, and Characterization of Food Proteins. *Foods* **2022**, *11*, 70. https://doi.org/10.3390/foods11010070

Received: 3 November 2021
Accepted: 15 December 2021
Published: 29 December 2021

One of the greatest challenges currently facing our society is combating climate change. The agricultural and food sectors are critical, as they are responsible for a considerable share of greenhouse gas emissions. Increasing competition for available resources emphasises the need to change consumption and food production styles. This is the only way to continue to have enough raw materials to produce food for a growing world population.

Reducing the production of animal proteins and replacing them with plant proteins could make an essential contribution to a more sustainable diet. For this reason, the recovery of functional proteins from plant raw materials or by-products of food production is currently an essential and strategic research topic in both food science and the food industry. Protein production is important, but so is the characterisation of the products' properties, such as their sensory or physicochemical behaviours. Only those protein products with good taste and high functionality will have long-term success in the market.

This Special Issue belongs to the section "Food Physics and (Bio)Chemistry", and is focused on the utilisation of protein-rich raw materials and by-products from the food industry to produce food ingredients and food products. It also shows how the functional properties of proteins can be improved. This Special Issue includes seven manuscripts, six being research papers and one a review paper, all of which are important contributions to this topic made by distinguished experts in this area.

Manuscript 1 is titled "Rejuvenated Brewer's Spent Grain: EverVita Ingredients as Game-Changers in Fibre-Enriched Bread" [1]. It describes the use of brewer's spent grain —the main side-stream of brewing—as a raw material for producing functional protein and dietary fibre-rich ingredients. The study investigated the impact of two Brewer's spent grain ingredients on dough and bread quality, and the nutritional value of the bread. The dietary fibre-rich ingredient had almost no influence on dough and bread quality. However, the protein-rich ingredient significantly improved the nutritional value of the bread by increasing the protein content by 36%. Thus, both ingredients could be game changers in developing bread fortified with dietary fibres and proteins.

Manuscript 2 ("Extrusion Processing of Rapeseed Press Cake-Starch Blends: Effect of Starch Type and Treatment Temperature on Protein, Fiber and Starch Solubility") [2] is a fascinating study on the behaviour of a protein-rich rapeseed press cake in a starch matrix during extrusion. In the extruded blends, the starch type significantly impacted protein solubility, which decreased with increasing barrel temperature. These effects could be attributed to the inconsistent process conditions, due to the different rheological properties of the starches, rather than to molecular interactions of the starches with the rapeseed proteins. Since texturates play an important role in the production of meat substitutes, these results can contribute important insights into the behaviour of the proteins in the end product, and possibly also to their nutritional evaluation.

Manuscript 3 "Screening of Twelve Pea (*Pisum sativum* L.) Cultivars and their Isolates Focusing on the Protein Characterization, Functionality, and Sensory Profiles" is dedicated to a raw material currently in great demand for the production of protein isolates and concentrates—peas (*Pisum sativum* L.) [3]. The authors investigated the suitability of twelve pea cultivars for the production of protein ingredients. While the protein yield differed between the cultivars, their proximate composition and physicochemical properties were quite similar. However, there were some differences in sensory properties in particular in the pea-like aroma attribute and the bitter taste.

Manuscript 4 entitled "Assessment of the Physicochemical and Conformational Changes of Ultrasound-driven Proteins Extracted from Soybean Okara Byproduct" is a valuable work about the utilization of ultrasound to extract proteins from okara—the by-product of soy food manufacturing [4]. Besides the protein yields and protein solubility, changes in the protein structure and an enhanced peptide generation were observed. Therefore, the okara protein extracted through the ultrasound technique is a valuable raw material for new applications.

Manuscript 5 ("Functional Properties of Rye Prolamin (Secalin) and their Improvement by Protein Lipophilization through Capric Acid Covalent Binding") deals with the modification of the prolamin fraction of rye protein to improve its functional properties [5]. Using lyophilization, the protein solubility and water absorption capacity decreased, but an increase in oil absorption, emulsifying/foaming capacity and stability was observed.

An improvement of the functional properties of lupin protein isolates using enzymatic hydrolysis followed by lactic fermentation is shown in Chapter 6, entitled "Fermentation of Lupin Protein Hydrolysates—Effects on their Functional Properties, Sensory Profile and the Allergenic Potential of the Major Lupin Allergen Lup an 1" [6]. In particular, the protein solubility and foaming activity at pH 4.0 increased significantly. In addition, Bead-Assay showed a successful reduction of the allergen fraction Lup an 1.

Manuscript 7 contains a review entitled "Barley Protein Properties, Extraction and Applications, with a Focus on Brewers Spent Grain Protein" [7]. This manuscript is the only review in this Special Issue/book, and excellently completes this book by giving a comprehensive overview of barley and brewers' spent grain proteins. The review explores the extraction and application of proteins from these raw materials, and offers insights into the protein composition. Brewers spent grains are a by-product of the brewery industry that is produced in large quantities. The utilisation of brewers' spent grains is thus an excellent example of the complete utilization of agricultural raw materials for sustainable nutrition.

The manuscripts in this book deal with the undeniable importance of protein ingredients for the food industry. At the same time, they show the diverse research opportunities that the recovery, isolation, and characterisation of food protein entail. Proteins can be extracted from plant materials with tailor-made functional properties. In addition, the utilisation of by-products of food production in food production offers the possibility for the holistic use of our agricultural products, with maximum exploitation of our rural resources. Both strategies—the increased cultivation and processing of protein-rich raw materials and the increased utilisation of agricultural products—offer promising ways to make our consumption and production styles more sustainable in the future.

Funding: This research received no external funding.

Conflicts of Interest: The authors declare no conflict of interest.

References

1. Sahin, A.W.; Atzler, J.J.; Valdeperez, D.; Münch, S.; Cattaneo, G.; O'Riordan, P.; Arendt, E.K. Rejuvenated Brewer's Spent Grain: EverVita Ingredients as Game-Changers in Fibre-Enriched Bread. *Foods* **2021**, *10*, 1162. [CrossRef] [PubMed]
2. Martin, A.; Naumann, S.; Osen, R.; Karbstein, H.P.; Emin, M.A. Extrusion Processing of Rapeseed Press Cake-Starch Blends: Effect of Starch Type and Treatment Temperature on Protein, Fiber and Starch Solubility. *Foods* **2021**, *10*, 1160. [CrossRef] [PubMed]
3. García Arteaga, V.; Kraus, S.; Schott, M.; Muranyi, I.; Schweiggert-Weisz, U.; Eisner, P. Screening of Twelve Pea (*Pisum sativum* L.) Cultivars and Their Isolates Focusing on the Protein Characterization, Functionality, and Sensory Profiles. *Foods* **2021**, *10*, 758. [CrossRef] [PubMed]
4. Aiello, G.; Pugliese, R.; Rueller, L.; Bollati, C.; Bartolomei, M.; Li, Y.; Robert, J.; Arnoldi, A.; Lammi, C. Assessment of the Physicochemical and Conformational Changes of Ultrasound-Driven Proteins Extracted from Soybean Okara Byproduct. *Foods* **2021**, *10*, 562. [CrossRef] [PubMed]

5. Qazanfarzadeh, Z.; Kadivar, M.; Shekarchizadeh, H.; Porta, R. Functional Properties of Rye Prolamin (Secalin) and Their Improvement by Protein Lipophilization through Capric Acid Covalent Binding. *Foods* **2021**, *10*, 515. [CrossRef] [PubMed]
6. Schlegel, K.; Lidzba, N.; Ueberham, E.; Eisner, P.; Schweiggert-Weisz, U. Fermentation of Lupin Protein Hydrolysates—Effects on Their Functional Properties, Sensory Profile and the Allergenic Potential of the Major Lupin Allergen Lup an 1. *Foods* **2021**, *10*, 281. [CrossRef] [PubMed]
7. Jaeger, A.; Zannini, E.; Sahin, A.W.; Arendt, E.K. Barley Protein Properties, Extraction and Applications, with a Focus on Brewers' Spent Grain Protein. *Foods* **2021**, *10*, 1389. [CrossRef]

 MDPI

Article

Rejuvenated Brewer's Spent Grain: EverVita Ingredients as Game-Changers in Fibre-Enriched Bread

Aylin W. Sahin [1], Jonas Joachim Atzler [1], Daniel Valdeperez [2,3], Steffen Münch [2,3], Giacomo Cattaneo [2], Patrick O'Riordan [3] and Elke K. Arendt [1,4,*]

1 School of Food and Nutritional Sciences, University College Cork, T12 YN60 Cork, Ireland; aylin.sahin@ucc.ie (A.W.S.); 117107223@umail.ucc.ie (J.J.A.)
2 EverGrain, LLC, One Busch Place, St. Louis, MO 63118, USA; Daniel.Valdeperez@Everingredients.com (D.V.); Steffen.Muench@Everingredients.com (S.M.); Giacomo.Cattaneo@everingredients.com (G.C.)
3 Global Innovation & Technology Centre, Anheuser-Busch InBev nv/sa, 3000 Leuven, Belgium; Patrick.ORiordan@ab-inbev.com
4 APC Microbiome Institute, University College Cork, T12 K8AF Cork, Ireland
* Correspondence: e.arendt@ucc.ie; Tel.: +353-21-4902064

Abstract: Brewer's spent grain (BSG) is the main side-stream of brewing. BSG is a potential source for nutritionally enriched cereal products due to its high content of fibre and protein. Two novel ingredients originating from BSG, EverVita FIBRA (EVF) and EverVita PRO (EVP), were incorporated into bread in two addition levels to achieve a 'source of fibre' (3 g/100 g) and a 'high in fibre' (6 g/100 g) nutrition claim for the breads. The impact of those two ingredients on dough and bread quality as well as on nutritional value was investigated and compared to baker's flour (C1) and wholemeal flour (C2) breads. The addition of EVF performed outstandingly well in the bread system achieving high specific volumes (3.72–4.66 mL/g), a soft crumb texture (4.77–9.03 N) and a crumb structure comparable with C1. Furthermore, EVF barely restricted gluten network development and did not influence dough rheology. EVP increased the dough resistance (+150%) compared to C1 which led to a lower specific volume (2.17–4.38 mL/g) and a harder crumb (6.25–36.36 N). However, EVP increased the nutritional value of the breads by increasing protein content (+36%) and protein quality by elevating the amount of indispensable amino acids. Furthermore, a decrease in predicted glycaemic index by 26% was achieved and microbial shelf life was extended by up to 3 days. Although both ingredients originated from the same BSG, their impact on bread characteristics and nutritional value varied. EVF and EVP can be considered as game-changers in the development of bread fortified with BSG, increasing nutritional value, and promoting sustainability.

Keywords: BSG; plant protein; fibre fortification; glycaemic index; bread dough quality; gluten network; nutritional value

Citation: Sahin, A.W.; Atzler, J.J.; Valdeperez, D.; Münch, S.; Cattaneo, G.; O'Riordan, P.; Arendt, E.K. Rejuvenated Brewer's Spent Grain: EverVita Ingredients as Game-Changers in Fibre-Enriched Bread. *Foods* **2021**, *10*, 1162. https://doi.org/10.3390/foods10061162

Academic Editors: Ute Schweiggert-Weisz and Emanuele Zannini

Received: 29 April 2021
Accepted: 20 May 2021
Published: 22 May 2021

Publisher's Note: MDPI stays neutral with regard to jurisdictional claims in published maps and institutional affiliations.

1. Introduction

Due to increased environmental awareness, the sustainable use of side-streams generated within the food industry has become increasingly important, focused on eliminating waste and the continual use of resources. Various studies have been conducted to valorise food waste by recycling these materials into biofuels, enzymes and bioactive compounds [1]. From a nutritional perspective, some of these by-products provide a rich source of nutrients, including vitamins, minerals, protein, fibre and bioactive compounds and, therefore, may be useful for further food applications [2–4]. Brewer's spent grain (BSG), the insoluble barley residue left after wort production, is the primary side-stream of brewing, representing 85% of the total by-products obtained [5]. The annual production is estimated to be 40 million tonnes worldwide, with 3.4 million tonnes generated in the European Union [6,7]. Currently, most BSG is used as a low-value animal feed with a cost of €35 per tonne [8].

BSG has huge potential to enhance the nutritional value of food due to its high content of fibre (30–50% *w/w*) and protein (19–30% *w/w*). The fibre is constituted of hemicellulose including arabinoxylans, cellulose and lignin [9]. The presence of arabinoxylans in the diet has been associated with various health benefits, including improved glycaemic control, reduced cholesterol, enhanced mineral absorption, faecal bulking and gut health [10,11]. The protein content of BSG is significant due to its composition of essential amino acids (~30%), especially lysine (~14.3%) [12].

The incorporation of BSG into cereal-based foods has been evaluated in several studies [12–16]. Bread is an important staple food worldwide due to its convenience, versatility, and low cost. In fact, bread has been reported as the main contributor to fibre intakes of Irish adults, although over 80% of the population are not reaching the European Food Safety Authority (EFSA) recommended intake of 25 g/d [17,18]. Therefore, the fibre enrichment of bread with BSG could be an adequate method to close the fibre gap. Several studies revealed the increase in the nutritional value of bread fortified with BSG showing an elevated protein and fibre content with high lysine concentrations and lower glycaemic index [19,20]. However, the replacement of wheat flour with BSG negatively impacts bread quality, resulting in a product with a lower volume, increased hardness, denser structure and a darker colour [12,16,21]. The addition of sourdough as a functional ingredient showed a positive impact on bread quality when BSG was added [22].

EverGrain (St. Louis, MO, USA) processed BSG and launched two ingredients, EverVita FIBRA and EverVita PRO, which differ in their protein and fibre content and particle size. A recent study characterised those ingredients and investigated their effect on pasta, resulting in outstanding product quality and increased nutritional value [23]. The current study reveals the impact of those two ingredients in two addition levels on bread dough quality including their effect on gluten network formation and starch pasting properties, as well as on final bread quality. Moreover, their effect on starch digestibility was determined using an in vitro starch digestibility model, and the protein quality in the bread based on the amount of indispensable amino acids was evaluated. A correlation analysis was conducted to provide a deep insight into the connected parameters, and two-way analysis of variance (ANOVA) revealed the influence of addition level or type of ingredient added and emphasised interaction effects of those two variables.

2. Materials and Methods

2.1. Raw Materials

Baker's flour and stoneground wholemeal flour supplied by Odlums (Dublin, Ireland), sunflower oil (Musgrave Wholesale Partners, Cork, Ireland), salt (Glacia British Salt Limited, Cheshire, UK), sugar (Nordzucker Ireland Ltd., Dublin, Ireland), baker's instant active dried yeast (Bruggemean Puratos, Dilbeek, Belgium) and tap water were used in the bread production. The two ingredients, EverVita FIBRA (EVF) and EverVita PRO (EVP), rejuvenated brewer's spent grain, were supplied by EverGrain, LLC (St. Louis, MO, USA). Chemicals used for analysis were purchased from Sigma-Aldrich (St. Louis, MO, USA), unless stated otherwise.

2.2. Nutritional Profile and Amino Acid Composition of Raw Ingredients

The compositional analysis of EVF and EVP were previously reported by Sahin et al., (2021) [23]. Baker's flour and wholemeal flour were analysed by Concept Life Sciences Ltd., Manchester, UK using the following methods: moisture—gravimetric air-oven method at 130 °C; protein—modified Dumas method with nitrogen-to-protein conversion factor 6.25 (AOAC 197.992.15); fat—low resolution proton nuclear magnetic resonance (NMR) (MQC-23-35; Oxford instruments application note); dietary fibre—gravimetric method (AOAC 991.43); ash—removal of organic matter by oxidation (550 °C) (ISO 936:1998); sodium—flame photometry; carbohydrates—calculated by difference. The sugar profile including fructose, glucose, sucrose/maltose and maltotriose/raffinose were determined using a Dionex ICS-5000+ system (Thermo Fisher Scientific, Sunnyvale, CA, USA) equipped with an electrochemical detector. Samples were extracted in triplicate following the procedure reported by Hoehnel et al. (2020) [24] and analyses on the ICS-5000+ system were performed as stated by Ispiryan et al. (2019) [25].

Amino acid analysis was conducted by Mérieux NutriScience CHELAB S.r.l., Resana, Italy. After protein extraction and hydrolysis (independent hydrolysis procedure for analysis of tryptophan, sulphuric amino acids and remaining amino acids), amino acids were quantified using ionic chromatography with post-column ninhydrin derivatisation coupled with a fluorescence and an ultraviolet (UV) detector. The concentration of the amino acids is given as % based on protein.

2.3. Bread Dough Preparation Process

Bread doughs were produced using the ingredients according to the recipes illustrated in Table 1. Two different concentrations EverVita FIBRA and EverVita PRO were incorporated into a bread formulation by replacing baker's flour. The addition levels were chosen to reach a 'source of fibre' (SF) and a 'high in fibre' (HF) claim according to European Union (EU) regulation [26].

Table 1. Recipes of breads in % based on baker's flour/wholemeal flour + EverVita ingredient.

	Controls		Source of Fibre		High in Fibre	
Ingredients	**Baker's Flour (C1)**	**Wholemeal Flour (C2)**	**EverVita FIBRA**	**EverVita PRO**	**EverVita FIBRA**	**EverVita PRO**
Bakers flour	100.00	n.a.	96.00	95.00	89.00	84.00
Wholemeal flour	n.a.	100.00	n.a.	n.a.	n.a.	n.a.
EverVita ingredient	n.a.	n.a.	4.00	5.00	11.00	16.00
Sunflower oil	3.20	3.20	3.20	3.20	3.20	3.20
Salt	1.20	1.20	1.20	1.20	1.20	1.20
Sugar	2.00	2.00	2.00	2.00	2.00	2.00
Bakers yeast	2.00	2.00	2.00	2.00	2.00	2.00
Water (25 °C)	60.47	63.30	62.27	63.97	64.30	70.73

n.a. represents 'not applicable'

Dry ingredients were pre-mixed for 1 min at minimum speed using a Kenwood Titanium Major KM020 mixer equipped with a hook attachment (Kenwood, Havant, UK) to ensure a homogeneous distribution, and instant active dried yeast was activated by its addition to water (25 °C) for 10 min. Yeast solution and sunflower oil were added to the dry ingredients and mixed at speed 1 for 1 min first, followed by a second mixing step at speed 2 for 7 min. This procedure was used for the baking process as well as for the evaluation of the fermentation quality of the bread doughs.

2.4. Fundamental Understanding of the Effect of EverVita Ingredients on Dough Properties

EverVita ingredients originate from BSG, which has been reported to impact bread dough quality negatively. Hence the effect of reinvented BSG-based ingredients on gluten network development, starch pasting properties as well as on dough rheology, extensibility and fermentability was investigated. Therefore, blends of baker's flour and EVF/EVP according to the ratios given in Table 1 were analysed. As controls baker's flour (control 1) and wholemeal flour (control 2) were chosen.

2.4.1. Gluten Network Formation and Starch Pasting Properties

The impact of EverVita ingredients on both gluten network and starch pasting was determined by the GlutoPeak and the Rapid Visco Analyser (RVA). For both measurements the solid part represents a mixing of baker's flour and EverVita ingredient (EVF or EVP) in the proportion used for the bread baking process (Table 1).

The GlutoPeak® (Brabender GmbH and Co. KG, Duisburg, Germany) was used to measure the gluten aggregation properties represented by the time of full aggregation (peak maximum time (PMT) in seconds (s)) and gluten network strength (torque maximum (TM) in Brabender units (BU)). 9 g of solids, based on 14% moisture, was added to 9 g of distilled water (36 °C) and the test was started using a rotation speed of 2750 rpm of the paddle.

Starch pasting properties during heating and cooking were determined using a Rapid Visco Analyser (Newport Scientific, Warriewood, Australia); 3 g of solids, based on 14% moisture, was added to 25 g distilled water in the metal sample cup and premixed briefly with the paddle to remove lumps. A constant shear rate of 160 rpm was applied during the measurement. The temperature profile used was equilibration at 50 °C for 1 min, followed by an increase to 95 °C with a heating rate of 0.2 °C/s, held at 95 °C for 162 s, cooled to 50 °C with a cooling rate of 0.2 °C/s, and held at 50 °C for 60 s. During the measurement the viscosity in centipoise (cP) was monitored resulting in a viscosity curve over time and peak viscosity (cP), breakdown viscosity (cP), trough viscosity (cP) and final viscosity (cP) were evaluated.

2.4.2. Water Absorption and Dough Development during Mixing

The water content of the different formulations was adjusted using Farinograph-TS® equipped with an automatic water dosing unit (Brabender GmbH and Co KG, Duisburg, Germany). Therefore, flour and EverVita ingredients were used in the ratio illustrated in Table 1. The target consistency of the doughs was 500 farinograph units (FU). After titration the optimal water content was used to determine the water absorption (WA; amount of water required to achieve a dough consistency of 500 FU), the dough development time (min) (DDT; time required to reach maximum consistency), the arrival of dough stability (min) (S1; length of time that dough consistency is held at 500 FU during mixing), and the mixing tolerance index (FU) (MTI; torque difference between maximum torque and torque five minutes after maximum) were evaluated. The optimal water content was used for preparing bread doughs.

2.4.3. Extensibility and Resistance to Extension

The dough extensibility (E) in mm, resistance to extension (RE) in extensograph units (EU) and the ratio of resistance over extensibility (RE/E) in (EU/mm) were determined using Extensograph® (Brabender GmbH and Co KG, Duisburg, Germany). The dough was prepared following the bread dough preparation process omitting yeast. 150 ± 0.5 g of dough (without addition of yeast) was moulded and placed into the proofing chamber. Analysis was performed after 90 min of proofing (time used for proofing during the baking process).

2.4.4. Fermentation Quality

A rheofermentometer (Chopin, Villeneuve-la-Garenne, France) was used to measure gaseous release and dough rise during the proofing process. We prepared 300 g dough following the bread dough preparation process procedure mentioned before and transferred into the fermentation chamber (30 °C). A cylindrical weight of 1500 g was placed on the dough and the dough was fermented for 180 min. The maximum height of the dough (Hm) in mm, the time required to achieve maximum height (T1) in minutes, the total volume of carbon dioxide released by the dough (Vtot) in mL and the height of maximum gas formation (H'm) in mm were evaluated.

2.5. Baking Process

Dough was separated into 450 g pieces, moulded, transferred into a bread pan (dimensions: 15 × 9.5 × 9.7 cm) and placed into the proofing chamber (KOMA BV Sunriser, Roermond, The Netherlands) for 90 min set to 30 °C and 85% relative humidity. After proofing, the tins with the leavened doughs were transferred into a deck oven (MIWE Condo, Arnstein, Germany) (220 °C top temperature and 230 °C bottom temperature). 400 mL of steam was injected prior to loading. The breads were baked for 35 min, removed from the baking tins and left to cool for 120 min on a rack before analysis. Two breads per batch were analysed after baking and two loaves were packed in plastic bags and stored at 20 °C for five days to determine the staling rate. Three batches per bread type were evaluated.

2.6. Changes in Techno-Functional Properties of Bread Fortified with EverVita Ingredients

2.6.1. Specific Volume

The specific volume was measured with a 3D laser scan using a Volscan Profiler 300 (Stable Micro Systems, Godalming, UK) and expressed as mL/g.

2.6.2. Crumb Texture and Staling

The bread crumb texture was analysed using a TA-XT2i texture analyser (Stable Micro Systems, Godalming, UK) equipped with a 25 kg load cell. For analysis, the breads were sliced with a thickness of 25 mm. To imitate chewing activity, a two-compression test was carried out using a cylindric probe with a diameter of 35 mm, a test speed of 5 mm/s, a trigger force of 0.05 N and a compression of 40%. The crumb hardness in Newton (N) and chewiness (N) were evaluated 120 min after baking. Gumminess (N) showed the same values as chewiness and was not further considered in the manuscript. In addition, the hardness (N) was measured 120 h after baking to evaluate bread staling. Staling was expressed as staling rate as defined by Sahin, Axel, Zannini, and Arendt (2018) [27].

2.6.3. Crumb Macro- and Microstructure

For the evaluation of crumb structure, a C-cell Imaging System (Calibre Control International Ltd., Warrington, UK) was used and slice area (mm^2), number of cells and average cell diameter (mm) were evaluated. The crumb area for the evaluation of the number of cells and the cell diameter was the slice area. Furthermore, the microstructure of the crumb was visualised using a scanning electron microscope (SEM). Therefore, bread crumb was freeze-dried, immobilised, and coated with a layer of 25 mm of sputtered palladium-gold. The microstructure was observed using a working distance of 8 mm and micrographs were taken at an accelerating voltage of 5 kV. SEM Control User Interface software, Version 5.21 (JEOL Technics Ltd., Tokyo, Japan) was used.

2.6.4. Crust and Crumb Colour

A colorimeter (Minolta CR-331, Konica Minolta Holdings Inc., Osaka, Japan) was used to determine the bread colour. To evaluate the influence of EverVita ingredients on the colour and the differences in colour compared to both controls, baker's flour and wholemeal flour, the ΔE value was calculated using the equation:

$$\Delta E = \sqrt{(L_2^* - L_1^*)^2 + (a_2^* - a_1^*)^2 + (b_2^* - b_1^*)^2}$$

where:

L^*_2 = lightness value of the bread including EverVita ingredients; L^*_1 = lightness value of the control (baker's flour or wholemeal flour).

a^*_2 = redness value of the bread including EverVita ingredients; a^*_1 = redness value of the control (baker's flour or wholemeal flour).

b^*_2 = yellowness value of the bread including EverVita ingredients; b^*_1 = yellowness value of the control (baker's flour or wholemeal flour).

2.6.5. Water Activity and Microbial Shelf Life

The water activity was determined using a water activity metre (HygroLab, Rotronic, Bassersdorf, Switzerland). The microbial shelf life analysis of the breads were performed using the environmental challenge method as reported by Dal Bello et al. (2007) [28]. Briefly, breads were cut into 25 mm slices. Bread slices were exposed to the environment for 5 min on each side, followed by packing them separately in sterile plastic bag and heat sealing them. To ensure aerobic conditions and allow gas exchange, two filter pipettes were inserted in each bag. Mould growth was monitored for 14 days and visually evaluated, while samples were stored in a temperature-controlled room (20 ± 2 °C). Samples were rated as "mould-free", "<10% mouldy", "10–24% mouldy", "25–49% mouldy", ">50% mouldy".

2.7. Impact of the Addition of EverVita Ingredients on the Nutritional Value of Breads

2.7.1. Starch Composition, Predicted Protein and Fibre Content and Predicted Amino Acid Profile in Breads

Starch can be present in different forms, digestible and resistant towards digestion. The digestible starch and resistant starch content of freeze-dried bread samples were determined by an enzymatic method using the K-RAPRS kit (Megazyme, Bray, Ireland). The digestible and resistance starch content in fresh bread was calculated considering the moisture content. The sum of both starch components represents the total starch content. The predicted protein and fibre content were calculated based on nutritional information of the ingredients considering the bake loss. The expected indispensable amino acid (histidine, isoleucine, leucine, lysine, sulphur-containing amino acids (SAA), aromatic amino acids (AAA), threonine, tryptophan, valine) in the breads were calculated and are presented relative to the requirement pattern bread on an average intake of 0.66 g protein per kg composition in breads [29]. Therefore, the ratio of baker's flour and EVF/EVP as well as the amino acid profile of the single ingredients were taken into account as reported by Hoehnel et al. (2020) [24].

2.7.2. Predicted Glycaemic Index and Load

An in vitro digestion method, specific to fibre-enriched food, was carried out to determine the predicted glycaemic index (pGI) of the breads. The procedure was performed as described by Brennan and Tudorica (2008) [30] and 4 g of fresh bread crumb was used for analysis. The release of reducing sugars over time was measured spectrophotometrically and the pGI and the predicted glycaemic load (pGL) were calculated using the following formula:

$$\begin{aligned} GI = 105.52 \quad & \times \frac{\text{fibre}}{\text{digestible carbohydrates}} - 76.46 \times \frac{\text{protein}}{\text{digestible carbohydrates}} \\ & + 1.23 \times RSRI_{\text{at 150 min}} + 69.41 \times SDI_{\text{at 270 min}} - 83.87 \end{aligned} \tag{1}$$

Reducing sugars released (RSR) is calculated as:

$$RSR[\%] = \frac{A_{\text{sample}} * 500 * 0.95}{A_{\text{maltose}} * \text{carbohydrates}} \tag{2}$$

where A_{sample} refers to the absorbance at 546 nm of the sample treated with enzymes; 500 refers to the total volume of buffer; 0.95 is the conversion factor of starch to maltose by amylase; A_{maltose} refers to the absorbance of a 1 mg/mL maltose standard; digestible carbohydrates in 4 g sample expressed in mg.

The reducing sugar release index (RSRI) at 150 min is defined as:

$$RSRI = \frac{\text{RSR 150min (Sample)}}{\text{RSR 150min (Control)}} \times 100 \tag{3}$$

The 'control' refers to the baker's flour control.

Sugar diffusion index:

$$SDI = \frac{DIFF_{\text{maltose}}}{DIFF_{\text{sample+maltose}}} \tag{4}$$

where $DIFF_{maltose}$ is the diffusion of the maltose blank (1 g maltose in 50 mL buffer). The percentage of maltose able to diffuse through the tube in the presence of sample (DIFF):

$$DIFF = \frac{(A_{blank+maltose} - A_{blank}) \times 500}{A_{maltose} \times 200} \times 100 \quad (5)$$

where $A_{blank\ +maltose}$ refers to the absorbance of the sample blank with maltose addition; A_{blank} refers to the absorbance of the sample blank; 500 is the total volume of buffer; $A_{maltose}$ refers to the absorbance of the maltose blank; 200 is the weight of maltose in 'blank + maltose'-sample in mg.

2.8. Statistical Evaluation

All tests were carried out in triplicate, unless stated otherwise. A variance analysis (one-way ANOVA, $\alpha \leq 0.05$, Tukey test) was performed using Minitab 17. In addition, a two-way ANOVA was conducted to evaluate the influence of the type of ingredients and the addition level. Correlation analysis was performed using Microsoft Excel.

3. Results

3.1. Nutritional Composition of Raw Ingredients

The nutritional composition of the raw ingredients is an influencing factor of the final nutritional profile of the food product. The differences between baker's flour (C1), wholemeal flour (C2) and EverVita ingredients regarding moisture, protein, fat, total carbohydrates, ash and sodium are illustrated in Figure 1A. EverVita ingredients are low in moisture (<2%) due to the processing (drying) of their original material BSG. Moreover, EverVita ingredients are exceptionally high in protein and fibre. EVF is by 81% and 105% higher in protein compared to C1 and C2, respectively. EVP contains an even greater amount, which is 185% and 222% of the protein content of C1 and C2, respectively. Both ingredients are similar in fat content (EVF: 4.7 g/100 g; EVP 5.8 g/100 g) and ash (EVF: 4.3 g/100; EVP: 3.3 g/100 g) which are higher compared to C1 and C2. The sodium level in EVF and EVP was lower than in C1 and C2.

The total carbohydrate concentration in the EverVita ingredients is 6–19% lower compared to the flours, yet their dietary fibre content, illustrated in Figure 1B, is significantly higher. EVF has a total dietary fibre content of 65.7 g/100 g, which is by 21-fold and 9-fold higher than C1 and C2, respectively. 1.9 g/100 g of the total dietary fibre in EVF is soluble fibre, a concentration comparable with baker's flour. EVP on the other hand contains less dietary fibre (46.5 g/100 g) than EVF, which is still significantly higher compared to C1 (3.3 g/100 g) and C2 (7.1 g/100 g). The soluble fibre content in EVP is 3 g/100 g, which is 2.7-fold and 4.2-fold the amount present in C1 and C2, respectively. Besides dietary fibre, mono-, di- and tri-saccharides are present in the raw ingredients (Figure 1C). Baker's flour showed the lowest total sugar content (0.702 g/100 g), followed by EVF (0.951 g/100 g). Wholemeal flour and EVP showed the highest sugar contents (1.12 g/100 g and 1.13 g/100 g, respectively). Both EverVita ingredients had a higher content of sucrose/maltose compared to C1 and C2.

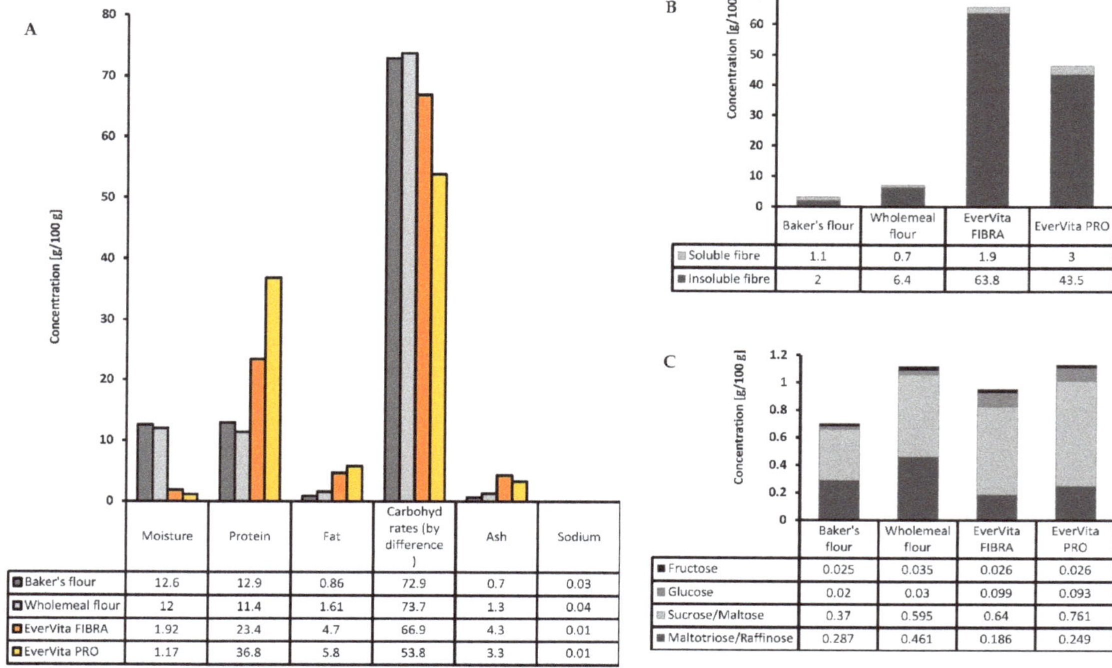

Figure 1. Composition of the raw ingredients baker's flour, wholemeal flour and EverVita ingredients, EverVita FIBRA and EverVita PRO (**A**), including the sugar profile (**B**) and the amount of soluble and insoluble dietary fibre (**C**). Values are expressed as mean values with a coefficient of variation < 0.1.

Apart from the macronutrients, such as protein, fat, and carbohydrates, the protein profile plays a key role in the final nutritional value of food products. The total amino acid composition of the protein fraction after hydrolysis of the raw ingredients based on their protein content is displayed in Figure 2. Fourteen of 18 amino acids showed their highest content in EVP among all raw ingredients. The concentrations were 1–75% higher compared to C1 and C2. A special emphasis needs to be put on the amount of indispensable amino acids which were all present in a higher amount in EVP compared to the other ingredients. In particular leucine and phenylalanine which make up 10.76% and 6.39% of the protein of EVP. Furthermore, both, EVP and EVF, contain tryptophan which was not detected in baker's flour or wholemeal flour.

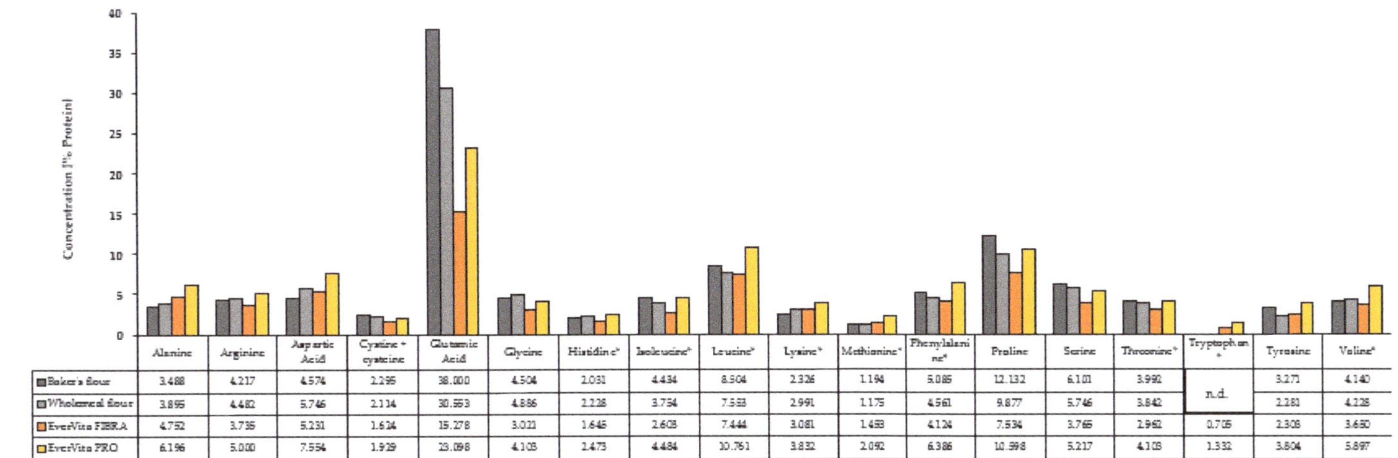

	Alanine	Arginine	Aspartic Acid	Cystine + cysteine	Glutamic Acid	Glycine	Histidine*	Isoleucine*	Leucine*	Lysine*	Methionine*	Phenylalanine*	Proline	Serine	Threonine*	Tryptophan*	Tyrosine	Valine*
Baker's flour	3.488	4.217	4.574	2.295	38.000	4.504	2.031	4.434	8.504	2.326	1.194	5.085	12.132	6.101	3.992	n.d.	3.271	4.140
Wholemeal flour	3.895	4.482	5.746	2.114	30.553	4.886	2.228	3.754	7.563	2.991	1.175	4.561	9.877	5.746	3.842		2.281	4.228
EverVita FIBRA	4.752	3.735	5.231	1.624	15.278	3.021	1.645	2.603	7.444	3.081	1.453	4.124	7.534	3.765	2.962	0.705	2.308	3.650
EverVita PRO	6.196	5.000	7.554	1.929	23.098	4.103	2.473	4.484	10.761	3.832	2.092	6.386	10.998	5.217	4.103	1.332	3.804	5.897

Figure 2. Amino acid composition of the raw ingredients baker's flour, wholemeal flour and EverVita ingredients, EverVita FIBRA and EverVita PRO, in % based on protein. Values are expressed as mean values with a coefficient of variation < 0.1. * Indispensable amino acids. n.d. stands for 'not detected'.

3.2. Impact of EverVita Ingredients on Gluten Network Formation and Starch Pasting

The effect of EverVita ingredients on gluten network development compared to the controls baker's flour (C1) and wholemeal flour (C2) is illustrated in Figure 3. C1 showed a gluten network development as commonly seen in refined wheat flour [31–33], characterised by an immediate increase in torque (flour hydration), followed by a plateau (colliding of gliadins and glutenins), and a further increase reaching a torque maximum (TM) at 68.00 ± 0.00 BU after 65.00 ± 0.00 s (PMT). C2 resulted in a different curve pattern with a significantly lower TM (27.67 ± 1.15 BU) and a significantly higher PMT (126.00 ± 7.55 s) compared to all other samples. The replacement of baker's flour by EverVita ingredients caused changes in gluten network development, particularly when EVP was used. Changes intensified with increasing addition level of the ingredients. The curve pattern of EVF inclusion appeared to be similar to C1 reaching lower TM-values of 65.5 ± 0.7 BU and 52.2 ± 2.1 BU, in SF and HF formulations respectively. EVF addition led to faster gluten network development with 60.5 ± 2.1 s in EVF (SF) and 60.5 ± 0.7 s in EVF (HF). The incorporation of EVP resulted in a significantly lower TM compared to C1, but addition level did not impact the result (55.7 ± 0.6 BU (SF); 55.7 ± 1.2 BU (HF), while the concentration significantly ($p < 0.05$) influenced the PMT (63.3 ± 2.3 s (SF); 29.0 ± 1.0 s (HF)).

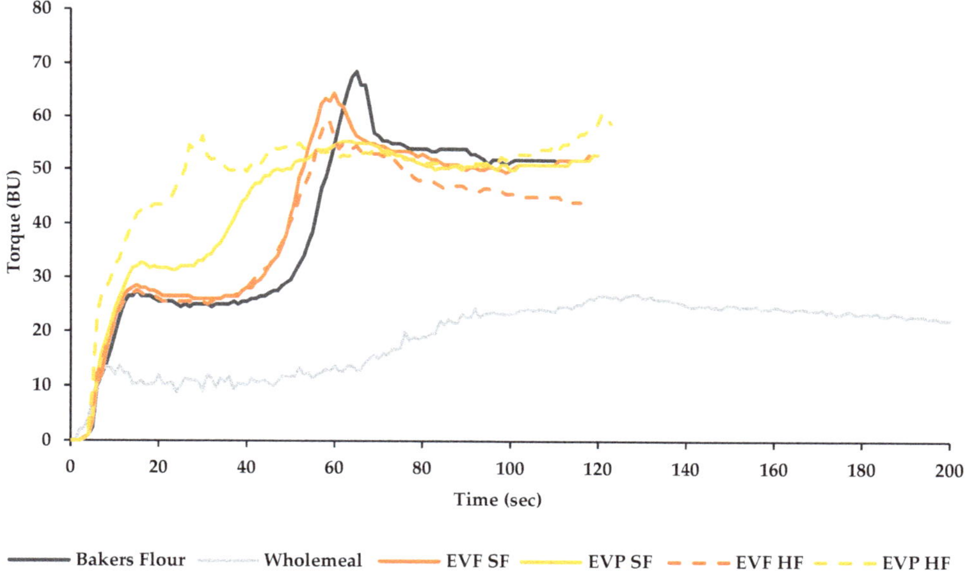

Figure 3. Gluten network development of baker's flour formulation enriched with EverVita ingredients (EverVita FIBRA (EVF), EverVita PRO (EVP)) at two inclusion levels (source of fibre (SF), high in fibre (HF)) compared to baker's flour control and wholemeal flour control. The curves represent average torque values of triplicates per sample. Fibre fortification using EverVita ingredients influenced the starch pasting properties as demonstrated in Table 2. Chosen decisive parameters are peak viscosity, breakdown viscosity, trough viscosity and final viscosity. The peak viscosity represents the maximum viscosity during heating and shearing and refers to the water binding capacity and swelling power of the starch. C1 showed the highest peak viscosity (1209 ± 9 cP) among the samples, while C2 showed the lowest (599 ± 33 cP). The incorporation of EverVita ingredients led to a significant lower degree of swelling compared to C1. This effect was advanced by HF addition levels of EVF (865 ± 2 cP) and EVP (812 ± 4 cP). After the peak viscosity, amylose and amylopectin leach out of the starch granules causing a decrease in viscosity to a certain trough. The trough viscosity indicates the holding strength of the system before retrogradation occurs.

Table 2. Pasting properties and dough analysis of samples. The results show mean values ± standard deviations. Values with the same lower-case letter in a row do not differ significantly from each other.

	Controls		Source of Fibre		High in Fibre	
	Bakers Flour (C1)	**Wholemeal Flour (C2)**	**EverVita FIBRA**	**EverVita PRO**	**EverVita FIBRA**	**EverVita PRO**
Peak Viscosity (cP)	1209 ± 9 (a)	599 ± 33 (e)	1077 ± 12 (b)	1067 ± 13 (b)	866 ± 2 (c)	812 ± 4 (d)
Breakdown (cP)	469 ± 4 (a)	101 ± 2 (d)	422 ± 10 (b)	422 ± 12 (b)	336 ± 8 (c)	347 ± 9 (c)
Trough (cP)	740 ± 6 (a)	516 ± 15 (c)	655 ± 2 (b)	645 ± 1 (b)	530 ± 10 (c)	465 ± 4 (d)
Final viscosity (cP)	1758 ± 9 (a)	1318 ± 28 (c)	1603 ± 8 (b)	1606 ± 1 (b)	1315 ± 8 (c)	1222 ± 11 (d)
Water absorption (%)	60.47 ± 0.15 (e)	63.30 ± 0.36 (c)	62.27 ± 0.25 (d)	63.97 ± 0.31 (bc)	64.30 ± 0.17 (b)	70.73 ± 0.25 (a)
Dough development time (min)	4.57 ± 0.28 (c)	10.12 ± 1.12 (b)	2.51 ± 0.37 (d)	2.48 ± 0.09 (d)	2.57 ± 0.02 (d)	17.84 ± 0.65 (a)
Dough stability at arrival (min)	1.42 ± 0.13 (b)	6.93 ± 0.32 (a)	1.33 ± 0.12 (b)	1.27 ± 0.12 (b)	1.50 ± 0.10 (b)	8.07 ± 0.98 (a)
Mixing tolerance index (FU)	34.33 ± 1.53 (a)	18.67 ± 6.35 (bc)	13.00 ± 2.83 (c)	19.67 ± 1.53 (b)	13.00 ± 4.00 (c)	Not detected
Extensibility (mm)	167.00 ± 2.94 (a)	63.00 ± 2.62 (f)	144.25 ± 6.90 (b)	131.75 ± 7.23 (c)	117.25 ± 2.06 (d)	79.00 ± 2.45 (e)
Resistance to extension (EU)	168.00 ± 9.63 (d)	363.25 ± 119.29 (bc)	260.00 ± 54.95 (cd)	350.25 ± 39.20 (bc)	425.50 ± 17.16 (b)	840.00 ± 36.00 (a)
Resistance to extension/Extensibility (EU/mm)	1.01 ± 0.07 (d)	5.75 ± 1.80 (b)	1.80 ± 0.33 (d)	2.66 ± 0.25 (cd)	3.63 ± 0.16 (c)	10.64 ± 0.57 (a)
Hm (mm)	77.50 ± 1.05 (a)	16.63 ± 1.96 (e)	68.80 ± 1.37 (b)	58.57 ± 2.50 (c)	53.70 ± 2.57 (c)	25.6 ± 1.4 (d)
T1 (min)	88.50 ± 6.36 (b)	71.00 ± 3.12 (b)	104.25 ± 3.18 (b)	172.50 ± 7.79 (a)	177.75 ± 3.18 (a)	175.5 ± 6.5 (a)
H'm (mm)	136.27 ± 3.22 (ab)	139.07 ± 2.51 (a)	138.47 ± 0.81 (ab)	139.07 ± 5.95 (a)	131.43 ± 3.76 (ab)	126.9 ± 6.3 (b)
Vtot (mL)	2601 ± 40 (a)	2498 ± 10 (ab)	2570 ± 46 (ab)	2585 ± 85 (a)	2476 ± 70 (ab)	2426 ± 48 (b)

C1 showed the highest trough viscosity (740 ± 6 cP), followed by formulations including EVF (SF) (655 ± 2 cP) and EVP (SF) (645 ± 1 cP). The incorporation of EVP (HF) showed a significant lower trough viscosity (465 ± 4 cP). The trough viscosity of C2 (516 ± 15 cP) was lower compared to C1 and did not significantly differ from samples including EVF (HF) (530 ± 10 cP). During cooling amylose and amylopectin reassociate, which is called retrogradation, and leads to an increase in viscosity. The highest final viscosity was determined in the C1 (1758 ± 9 cP), followed by the formulations with EverVita ingredients at a SF level (EVF (SF): 1603 ± 8 cP; EVP (SF): 1606 ± 1 cP). The addition of EVF at a higher level (HF) led to a significant lower value (1315 ± 8 cP) which is comparable to the final viscosity detected in C2 (1318 ± 28 cP). The incorporation of EVP at HF level resulted in the lowest final viscosity (1222 ± 11 cP).

3.3. Rheological Properties of Doughs and Fermentation Quality

Rheological characteristics of dough include consistency changes during mixing (Farinograph) as well as elasticity and resistance to extension during stretching of the dough (Extensograph).

The farinograph was not only used to adjust the water content (WA) of each dough reaching a final consistency of 500 ± 20 FU, rheological parameters such as dough development time (DDT), arrival of dough stability (S1) and the mixing tolerance index (MTI) were investigated. Figure 4 shows the Farinograph curves of the different formulations and the results of DDT, S1 and MTI are shown in Table 2.

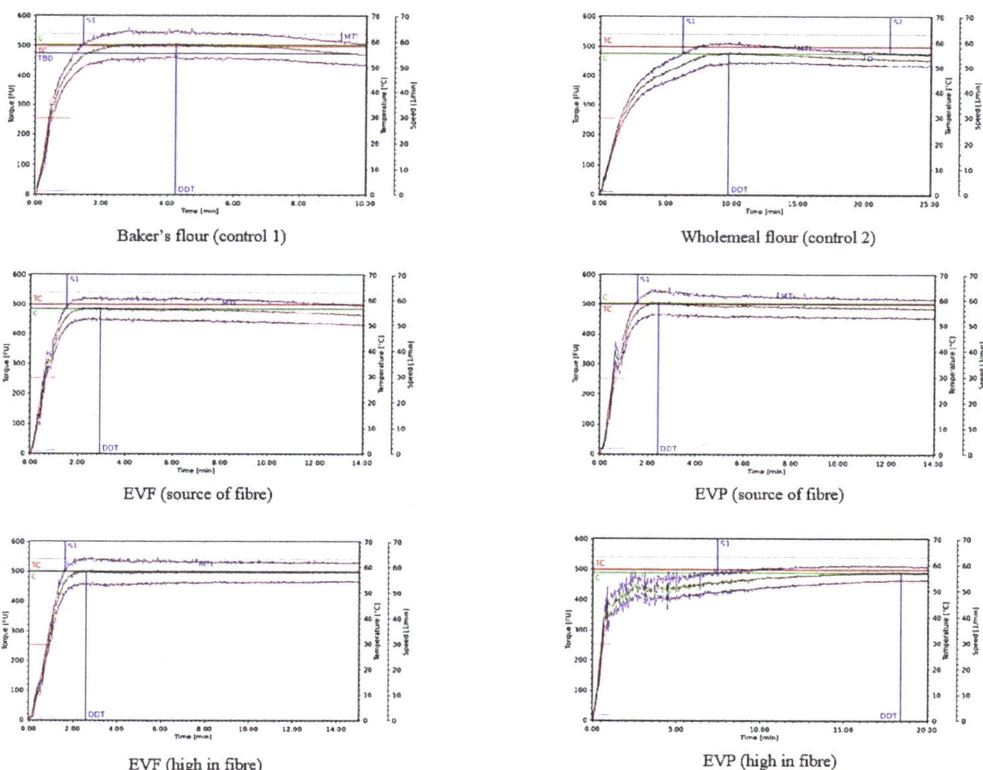

Figure 4. Farinograph curves of the controls (baker's flour and wholemeal flour) and the impact of EverVita FIBRA (EVF) and EverVita PRO (EVP) on dough rheology during mixing.

C1 showed the lowest water absorption resulting in a water addition level of 60.47% to achieve 500 FU, while C2 required 63.30% water. The addition of EVF (SF) and EVP (SF) caused an increased water addition level by 1.80% and 3.50% compared to C1, respectively. While the incorporation of EVF (HF) only affected the water absorption of the system to a relatively small extent (+3.83%) compared to C1, EVP (HF) caused an increase by 10.26% and showed the highest water content overall. DDT and S1 revealed a strong positive correlation ($p < 0.04$; $r = 0.95$). C2 resulted in a longer DDT (10.12 ± 0.28 min) and a delay in S1 (6.93 ± 0.32 min) compared to C1 (DDT: 4.57 ± 0.28; S1: 1.42 ± 0.13 min). Both EverVita ingredients caused a quicker DDT (EVF: 2.51 ± 0.37 min; EVP: 2.48 ± 0.09 min) and S1 is reached earlier when added in 'source of fibre' levels (EVF (SF): 1.33 ± 0.12 min; EVP (SF): 1.27 ± 0.12 min). While the addition level of EVF did not have any impact on the dough rheology during mixing, the longest overall DDT (17.84 ± 0.65 min) and the highest S1 (8.07 ± 0.98 min) occurred in doughs with EVP (HF). The MTI showed the highest value for C1 (34.33 ± 1.53) while all other doughs had a significantly lower MTI (between 13.00 to 19.67).

The Farinograph measurement of EVP (HF) did not give any MTI value which indicates no changes in peak consistency five minutes after the peak is reached. Furthermore, the curve showed fluctuations in the first 7.5 min of mixing.

Extensograph measurements revealed the extensibility (E) and the resistance to extension (RE) during stretching of the dough and the results are demonstrated in Table 2. C1 showed the highest extensibility (167.00 ± 2.94 EU), followed by EVF (SF) (144.25 ± 6.90 EU) and EVP (SF) (131.75 ± 7.23 EU). The lowest extensibility was detected in C2 (63.00 ± 2.94 EU) and EVP (HF) (79.00 ± 2.45 EU), while EVF (HF) resulted in a significantly higher dough extensibility (117.25 ± 2.06 EU). The RE value in C1 was the lowest (168.00 ± 9.63 mm), while EVP (HF) showed the highest RE (840.00 ± 36.00 mm). The ratio of RE over extensibility indicates the balance between dough strength and dough stretchability. The ratio in C1 was 1.01 ± 0.07. The addition of EverVita ingredients, especially EVP, increased the value resulting in the highest RE/E in EVP (HF) (10.64 ± 0.57 mm/EU).

The rheofermentometer was used to measure dough rise and CO2 formation during 180 min of fermentation. The results for maximum height of the dough (Hm), time required to achieve maximum height (T1), the total volume of carbon dioxide released by the dough (Vtot) and the height of maximum gas formation (H'm) are displayed in Table 2. C1 showed the greatest dough height during the leavening process (Hm = 77.50 ± 1.05 mm), while dough rise of C2 resulted in the lowest Hm value (16.63 ± 1.96 mm). The addition of EverVita ingredients caused a significantly lower Hm, especially when EVP was applied. This effect was amplified by a higher addition level of the ingredients leading to values of 53.70 ± 2.57 mm and 25.6 ± 1.4 mm in EVF (HF and EVP (HF), respectively. The time at which the maximum dough height was achieved was 88.50 ± 6.36 min in C1 which did not differ significantly from C2 (71.00 ± 3.12 min). The maximum gas formation did not differ significantly in the doughs except for EVP (HF) which showed a significantly lower H'm (126.9 ± 6.3 mm) compared to all other samples. This resulted in the same trend for Vtot with C1 (2601 ± 40 mL) and EVP (SF) (2585 ± 85 mL) showing the highest CO_2 production by yeast, while in dough including EVP (HF) the lowest CO_2 formation occurred (2426 ± 48 mL).

3.4. *Effect of EverVita FIBRA and EverVita PRO on Bread Quality*

The specific volume, crumb texture, crumb macro- and microstructure as well as the crust and crumb colour, water activity and microbial shelf life of the breads were determined to evaluate bread quality. The results are illustrated in Table 3.

Table 3. Technological properties of the bread samples. The results show mean values ± standard deviations. Values with the same lower-case letter in a row do not differ significantly from each other.

	Controls		Source of Fibre		High in Fibre	
	Bakers Flour (C1)	**Wholemeal Flour (C2)**	**EverVita FIBRA**	**EverVita PRO**	**EverVita FIBRA**	**EverVita PRO**
Specific Volume (mL/g)	4.46 ± 0.26 (a)	2.28 ± 0.07 (c)	4.66 ± 0.23 (a)	4.38 ± 0.31 (a)	3.72 ± 0.37 (b)	2.17 ± 0.05 (c)
Hardness (N)	4.76 ± 1.20 (e)	24.54 ± 3.68 (b)	4.77 ± 0.65 (e)	6.25 ± 1.49 (de)	9.03 ± 1.28 (cd)	36.36 ± 1.99 (a)
Chewiness (N)	8.39 ± 1.67 (c)	14.24 ± 1.79 (b)	3.25 ± 0.56 (f)	4.32 ± 1.26 (e)	6.40 ± 1.18 (d)	20.32 ± 1.20 (a)
Staling rate (-)	1.60 ± 0.31 (b)	1.32 ± 0.45 (bc)	2.66 ± 0.42 (a)	2.70 ± 0.51 (a)	2.55 ± 0.49 (a)	1.24 ± 0.15 (c)
Slice Area (mm^2)	10323 ± 590 (a)	4990 ± 388 (e)	10469 ± 432 (a)	9677 ± 640 (b)	8970 ± 568 (c)	5701 ± 369 (d)
Number of Cells	5228 ± 349 (b)	2794 ± 144 (c)	5434 ± 383 (b)	5997 ± 268 (a)	5234 ± 296 (b)	5336 ± 514 (b)
Cell Diameter (mm)	2.43 ± 0.20 (a)	1.25 ± 0.20 (d)	2.32 ± 0.12 (a)	1.88 ± 0.10 (c)	2.08 ± 0.11 (b)	1.31 ± 0.13 (d)
ΔE crust (compared to C1)	-	11.17 ± 2.65 (a)	10.91 ± 2.11 (ab)	10.95 ± 1.87 (ab)	9.03 ± 1.93 (ab)	8.21 ± 4.47 (b)
ΔE crust (compared to C2)	11.17 ± 2.65 (b)	-	15.08 ± 2.42 (a)	16.54 ± 2.55 (a)	11.27 ± 1.46 (b)	8.93 ± 2.30 (b)
ΔE crumb (compared to C1)	-	17.66 ± 2.86 (b)	5.28 ± 3.39 (c)	8.92 ± 2.80 (bc)	13.05 ± 3.74 (b)	22.43 ± 2.71 (a)
ΔE crumb (compared to C2)	17.66 ± 2.86 (a)	-	14.74 ± 3.43 (ab)	9.60 ± 2.63 (b)	6.14 ± 2.52 (b)	5.69 ± 2.20 (b)
Water activity	0.951 ± 0.008 (a)	0.953 ± 0.009 (a)	0.938 ± 0.019 (a)	0.947 ± 0.023 (a)	0.940 ± 0.026 (a)	0.947 ± 0.019 (a)
Day of first microbial growth	6.00 ± 1.00 (b)	6.67 ± 0.58 (b)	6.00 ± 0.00 (b)	9.00 ± 1.00 (a)	6.33 ± 0.58 (b)	9.33 ± 0.58 (a)

The addition of EverVita ingredients in a concentration to achieve a 'source of fibre' claim did not significantly impact the specific volume (EVF (SF): 4.66 ± 0.23 mL/g; EVP (SF): 4.38 ± 0.31 mL/g) compared to C1 (4.46 ± 0.26 mL/g). However, a higher addition level decreased the specific volume of the breads significantly resulting in 3.72 ± 0.37 mL/g and 2.17 ± 0.05 mL/g for EVF (HF) and EVP (HF) breads, respectively. EVP (HF) showed the same specific volume as C2 (2.28 ± 0.07 mL/g). In Figure 5 the differences in volume are visualised.

The texture of the bread crumb was evaluated considering crumb hardness (N) and chewiness (N), and the degree of staling over time was determined and expressed as the staling rate. The softest crumb was determined in C1 (4.76 ± 1.20 N) and EVF (SF) (4.77 ± 0.65 N), followed by EVP (SF) (6.25 ± 1.49 N) and EVF (HF) (9.03 ± 1.23 N). C2 showed a significant harder crumb (24.54 ± 3.68 N) and the highest crumb hardness was measured in bread containing EVP (HF) (36.36 ± 1.99 N). EVF (SF) resulted in the least chewy crumb (3.25 ± 0.56 N), while EVP (HF) had the highest chewiness value (20.32 ± 1.20 N). The staling rate was the highest in EVF (SF), EVP (SF) and EVF (HF) with values between 2.55 and 2.70. The lowest staling rates were detected in C2 (1.32 ± 0.45) and EVP (HF) (1.24 ± 0.15).

Changes in crumb structure were evaluated considering slice area (mm^2), number of cells and average cell diameter (mm). A strong positive correlation between slice area and specific volume ($p < 0.003$; $r = 0.99$) occurred and thus the biggest slice area was measured in C1 (10,323 ± 590 mm^2) and EVF (SF) (10,469 ± 432 mm^2) breads, while the smallest slice area was determined in C2 (4990 ± 388 mm^2) and EVP (HF) (5701 ± 369 mm^2). The highest number of cells in the bread crumb was determined in EVP (SF) (5997 ± 268). These cells were relatively small (diameter of 1.88 ± 0.10 mm) compared to those of the other samples. C2 showed the lowest number of cells (2794 ± 144) and also the smallest average cell diameter (1.25 ± 0.20 mm). C1 and EVF (SF) had the biggest cell diameter among all samples with 2.43 ± 0.20 mm and 2.32 ± 0.12 mm for C1 and EVF (SF), respectively. Differences in crumb structure were visualised using a SEM and are illustrated in Figure 5. The inclusion of EVF resulted in a crumb structure very similar to C1, while EVP caused a compact and dense crumb structure with a film covering the starch granules. Furthermore, the crumb surface occurred to be bigger overall in breads with EVF compared to EVP, in particular with high in fibre inclusion levels.

The crust and crumb colour of the breads were evaluated by determining the ΔE-value which indicated the difference in L*-, a*- and b*-value compared to C1 and C2 as controls. Compared to C1, C2 showed the biggest difference in crust colour (11.17 ± 2.65) and crumb colour (17.66 ± 2.86). The most similar crust colour to C1 was determined in breads including EVP (HF) (8.21 ± 4.47), while EVF (SF) (5.28 ± 3.39) and EVP (SF) (8.92 ± 2.80) showed the most similar crumb colour to C1. Compared to C2, breads containing EverVita ingredients with source of fibre addition level (EVF (SF): 15.08 ± 2.42; EVP (SF): 16.54 ± 2.55) had the highest ΔE-value, while EVP (HF) showed the most similar crust colour (8.93 ± 2.30). Regarding the colour difference in crumbs, EVP (HF) had the most similar crumb colour to C2 (5.69 ± 2.20), whereas C1 showed the highest difference.

The microbial shelf life of the breads revealed the shortest shelf life in C1 and EVF (SF) breads. Compared to C1 which showed the first microbial growth after 6.00 ± 1.00 days, the growth started after 9.00 ± 1.00 days and 9.33 ± 0.58 days in breads including EVP (SF) and EVP (HF), respectively. Hence, an extension of microbial shelf life occurred, even though no significant differences in water activity was determined.

Figure 5. Appearance of fresh bread samples and micrographs of the bread crumbs. The first row illustrates the controls, baker's flour and wholemeal flour; Breads including EverVita FIBRA and EverVita PRO in source of fibre level and in high in fibre level are demonstrated in the second and third row, respectively.

3.5. Modification of Nutritional Value of Breads

The nutritional value of breads was evaluated considering the composition of the breads (starch composition, protein and fat content, moisture content) as well as the amino acids composition and the predicted glycaemic index (pGI) and predicted glycaemic load (pGL).

The predicted composition of the breads based on ingredient characteristics and the addition level of EverVita ingredients is illustrated in Table 4.

Table 4. Starch, dietary fibre, protein, fat and moisture content of bread samples. Values represent the means ± standard deviation.

	Controls		Source of Fibre		High in Fibre	
	Baker's Flour (C1)	Wholemeal Flour (C2)	EverVita FIBRA	EverVita PRO	EverVita FIBRA	EverVita PRO
Total starch (g/100 g)	40.56 ± 0.42 (a)	34.00 ± 1.10 (bc)	39.26 ± 1.18 (ab)	36.23 ± 2.19 (abc)	35.23 ± 1.05 (abc)	31.66 ± 2.16 (c)
Of which is digestible starch (g/100 g)	39.82 ± 0.02 (a)	33.17 ± 1.12 (bc)	38.63 ± 1.21 (ab)	35.51 ± 2.22 (abc)	34.51 ± 1.06 (abc)	31.03 ± 2.23 (c)
Of which is resistant starch (g/100 g)	0.74 ± 0.44 (a)	0.82 ± 0.02 (a)	0.63 ± 0.02 (a)	0.72 ± 0.03 (a)	0.72 ± 0.03 (a)	0.63 ± 0.07 (a)
Fibre (g/100 g)	2.1 *	4.8 *	3.8 *	3.5 *	6.6 *	6.5 *
Protein * (g/100 g)	9.5 *	8.4 *	9.7 *	10.1 *	10.1 *	11.5 *
Fat (g/100 g)	3.18 *	3.33 *	3.24 *	3.27 *	3.36 *	3.47 *
Moisture (g/100 g)	43.17 ± 0.19 (bc)	44.37 ± 0.80 (b)	42.04 ± 0.04 (c)	43.85 ± 0.12 (b)	43.20 ± 0.49 (bc)	46.68 ± 0.26 (a)

(*) based on calculation considering the information from the compositional analysis of the raw ingredients. Values with the same lower-case letter in a row do not differ significantly from each other.

The main compound of the breads is moisture, which ranges between 42.04 ± 0.04% in EVF (SF) breads to 46.68 ± 0.26% in EVP (HF) breads. The second main compound in the breads is starch. The highest total starch content was determined in C1 (40.56 ± 0.42 g/100 g), whereas high fibre breads and C2 showed the lowest total starch concentrations. The same trend occurred in the digestible starch content. C1 had the lowest dietary fibre content (2.1 g/100 g) and C2 included 4.8 g/100 g dietary fibre based on calculation. EverVita ingredients were added in amounts needed to achieve either 3 g/100 g of dietary fibre ('source of fibre' claim) or 6 g/100 g ('high in fibre' claim). The inclusion of EverVita ingredients increased the protein content in the breads, which ranged between 8.4 g/100 g (C2) and 11.5 g/100 g (EVP (HF)). No major differences occurred in the fat content of the breads. C1 had the lowest fat content (3.18 g/100 g) and EVP (HF) breads included the highest fat content (3.47 g/100 g).

The amount of indispensable amino acids was calculated and expressed relative to the requirement pattern established by the World Health Organisation (WHO) (Figure 6). Since the inclusion level of 'source of fibre' breads including EverVita ingredients are relatively low, changes in amino acid profile were expected rather for 'high in fibre' breads; hence, source of fibre breads were neglected in the evaluation of the predicted amino acid composition of the final breads. The replacement of baker's flour by any of the EverVita ingredients did not result in an inferior amino acid score.

By contrast, the fortification of wheat bread with EVP increased the concentration of some indispensable amino acids, in particular lysine (+24.5%) an amino acid that is known to be limiting in cereal based products. Moreover, the incorporation of EVP increased the concentration of aromatic amino acids (AAA) by 4% compared to C1, and valine (+1%). A significant difference occurred in the predicted tryptophan content in both high in fibre breads (EVF and EVP), which made 0.215% and 0.561% based on the total protein content in the breads, respectively, while no tryptophan was expected in both controls.

The pGI and pGL values represent the in vitro starch digestibility of the breads during digestion. The pGI and pGL of the different bread samples are illustrated in Figure 7. C1 showed a pGI value of 87.49 ± 9.31, which was significantly higher than C2 with a pGI of 46.75 ± 0.20. The replacement of baker's flour by EverVita ingredients aiming for a 'source of fibre' claim according to EU regulations did not result in an inferior bread quality regarding starch digestibility. On the opposite: EVP (HF) breads caused a significant reduction in pGI (64.25 ± 2.53). The same trend was observed for the pGL values, with C2 (7.75 ± 0.03) and EVP (HF) (9.97 ± 0.39), which can be considered as low.

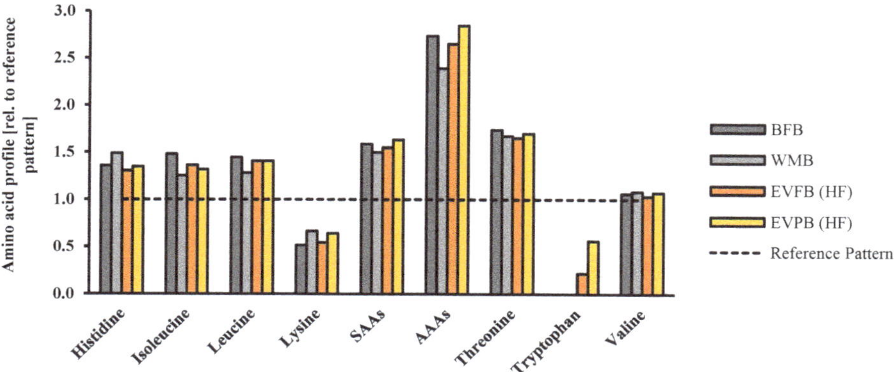

Figure 6. Profile of indispensable amino acids of the final breads baker's flour control (BFB), wholemeal flour control (WMB), high in fibre bread including EverVita FIBRA (EVFB (HF)) and high in fibre bread with EverVita PRO addition (EVPB (HF)). The values are expressed relative to the requirement pattern established by the World Health Organisation [29].

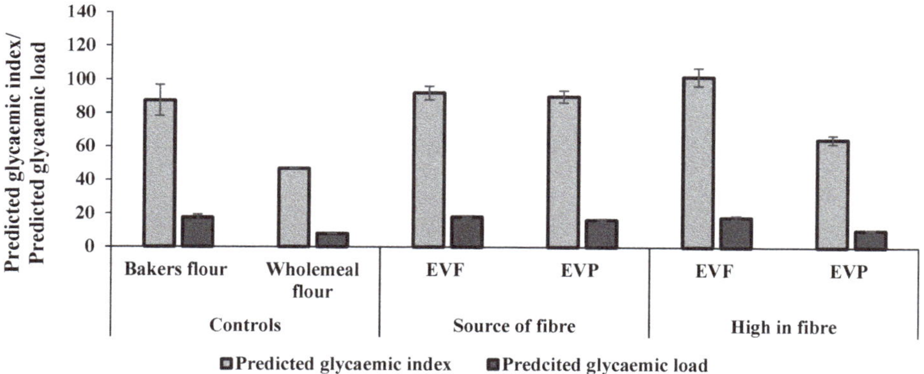

Figure 7. The predicted glycaemic index/glycaemic load of bread samples. EVF refers to EverVita FIBRA, while EVP represents EverVita PRO. SF and HF indicate the addition level of EverVita ingredients at 'source of fibre' and 'high in fibre', respectively.

4. Discussion

The impact of BSG on bread dough characteristics and final bread quality has been extensively studied. Rejuvenated BSG in the form of two ingredients, EverVita FIBRA and EverVita PRO, was used to fortify bread in two inclusion levels, 3 g/100 g and 6 g/100 g bread. The type of EverVita ingredient significantly impacted on dough and bread quality and final bread characteristics.

EVF contains 23.4% protein and 67.6% dietary fibre, mainly insoluble fibre, while EVP has a protein content of 36.8% and a dietary fibre content of 46.8% [23]. Both ingredients weakened the gluten network. The inclusion level was not the only parameter impacting network formation ($p < 0.001$), the type of EverVita ingredient also influenced the network strength ($p < 0.005$) and development time ($p < 0.001$). As previously reported, the EverVita ingredients contain low molecular weight peptides, which promote intramolecular connections, such as disulphide-, hydrogen- and ionic bonds resulting in a faster network development [23,34]. Since EVP has a higher protein content compared to EVF, it shortened the development time more effectively. EVF, on the other hand, influenced the network strength somewhat. Compared to EVP, EVF is higher in dietary fibre, that is mainly insoluble, which causes firstly changes in the secondary structure of gluten proteins due to the formation of hydrogen bonds between β-turns and the fibre, causing a weaker network [35,36]. Secondly, fibre competes with gluten for water leading to a lower degree of gluten hydration and hence a restricted network development [37]. Also, a physical hindrance of the gluten network development may occur, which was detected to a higher extent in wholemeal flour (C2) due to the presence of bran particles [38,39].

The gluten network strength influenced the dough rheology resulting in a positive correlation with extensibility ($p < 0.036$; $r = 0.84$) and dough rise during proving ($p < 0.039$; $r = 0.84$). This indicates the primary role of the gluten network on dough quality and the weakening effect caused by fibre ingredients [40–43]. Bread dough is considered ideal in its viscoelastic properties if the ratio of viscous and elastic parts is balanced, meaning RE/E is 1 [44]. Compared to EVF, EVP increased the resistance to extension enormously. Thus, the dough's elastic part increased, causing a higher dough stiffness, most likely due to the higher protein content. Proteins can form covalent bonds with other proteins and peptides [45] which increases the resistance of the system to external stress [46,47], such as mixing, for example, which is illustrated by the mixing tolerance index. This index indicates that formulations including EVP withstand the mixing to a greater extent than all other formulations.

The dough rise during fermentation (Hm) is influenced by both, addition level ($p < 0.0001$) and the type of EverVita ingredient ($p < 0.0001$) and showed a strong positive correlation with the gluten network strength ($p < 0.038$; $r = 0.84$). Baker's flour is rich in intact gluten proteins and starch, which participate in dough structure formation [48]. The replacement of flour by EVP or EVF caused a reduction of both intact gluten and starch in the dough system and reduced its viscoelastic behaviour. The replacement by EVF increased the amount of insoluble fibre in the system interacting with the gluten network. This weakened the network mainly due to physical interaction and lowered DDT [49,50]. On the other hand, the substitution by EVP caused a lower Hm due to an increase in resistance to extension of the dough proven by a strong positive correlation with RE/E ($p < 0.04$; $r = -0.84$). Furthermore, the curves of the gluten network development (Figure 3) illustrate a very high resistance of formulations including EVP reflected by no sudden breakdown of the network, which was also found in bread systems fortified with plant-based proteins [31]. Moreover, the smaller particle size allowed a higher degree of interaction with compounds, such as water, fibre, and other proteins [51]. It also needs to be mentioned that the addition level of EVP to reach the fibre claims was higher compared to EVF, which led to a reduction of wheat flour and related gluten concentration in the final system. This reduction also influenced the colour formation during baking which was mainly affected by the colour of the ingredients EVF and EVP.

Vtot is the total volume of CO_2 produced by yeast during proving, mainly influenced by the amount of available carbohydrates as a nutritional substrate for yeast. The replacement of baker's flour, which contains over 70% carbohydrates, mainly starch, by ingredients containing less than 5% starch [23], resulted in less available carbohydrates to be metabolised by yeast, and hence less CO_2 production. Thus, Vtot is only affected by the flour replacement level by EverVita ingredients ($p < 0.01$).

It is known that dough extensibility and dough expansion during proofing influence the final bread quality. Hm and RE/E correlated with several bread quality parameters, such as specific volume ($p < 0.003$; $r = 0.96$), crumb hardness ($p < 0.018$; $r = -0.89$), chewiness ($p < 0.05$; $r = -0.83$) and slice area ($p < 0.0005$; $r = 0.98$). The specific volume correlated negatively with the dough development time during mixing ($p < 0.02$; $r = -0.88$). To achieve the highest specific volume the formulation can reach, the dough needs to be fully developed, meaning all compounds need to be fully hydrated to form a network [52].

EVP (HF) showed a significantly long DDT, putatively due to the incorporation of high amounts of low molecular weight proteins competing with other compounds for water and hence delaying the hydration of the system [53]. This indicates that the mixing time used to prepare the bread dough might have been insufficient for all formulations resulting in a low specific volume [54]. EVP caused a significant reduction in specific volume, which led to a denser crumb structure, resulting in a harder crumb and a higher crumb chewiness, which both correlated negatively with the specific volume ($p < 0.01$; $r = 0.92$).

Besides the dough rheology, starch pasting during baking and the related changes in viscosity of the system impacts the final product quality. In general, the replacement of baker's flour caused lower viscosity values which may be related to the reduced amount of total starch. The peak viscosity showed a strong positive correlation with the specific volume of the bread ($p < 0.02$; $r = 0.89$), meaning a high degree of starch swelling during heating results in a high viscosity which enhances the stability of the system during the baking process [55], and leads to a high specific volume. The incorporation of EVP resulted in a lower peak viscosity due to the higher protein content and the related higher water absorption than EVF, causing an increasing competition with starch for water and a restricted starch swelling [56].

As already mentioned, EVP (HF) caused a significant decrease in the specific volume of the breads, which affected the crumb structure. Food structure affects the starch digestibility of products making starch less accessible for digestive enzymes to bind on the substrate, in this case, starch [57]. In particular, proteins are known to form a matrix in which starch granules are embedded. This appears to be a barrier towards enzymatic starch degradation [57]. Also, during the baking process, changes in protein conformation may occur, which may promote the formation of disulphide bonds [58]. Thus, EVP showed a high impact on reducing the pGI-value and the pGL. The partial replacement of baker's flour by EVF did not decrease the pGI, but it also did not result in an inferior bread quality regarding starch digestibility compared to C1. EVF not having any impact on the pGI could have been firstly, by a lower protein content and thus a weaker protein matrix protecting the starch granules from enzymatic attack [59], and secondly, by the bigger particle size of the ingredients causing a higher degree of physical interruption of the protein-starch network [60]. Micrographs of the crumb (Figure 5) also demonstrate the higher overall surface in bread with EVF compared to EVP, which increases the chance of enzymatic degradation. EVP not only influenced the starch digestibility positively, the addition of EVP in 'high in fibre' level also elevated the amount of indispensable amino acids in the bread. This was especially pronounced for lysine and tryptophan, which are known to be limited in cereal-based products [61]. Tryptophan is present in barley in higher amounts compared to wheat [62]. Hence, by replacing baker's flour with BSG-derived ingredients, such as EVF and EVP, a natural tryptophan fortification occurs. This indispensable amino acid contributes to protein synthesis in the human body and is involved in other physiological mechanisms, such as the synthesis of serotonin and vitamin B_3 [62,63]. In addition, dietary tryptophan has a high therapeutic potential to treat multiple chronic diseases such as cardiovascular disease, depression and inflammatory bowel disease, among others [62]. Even though the requirements for those two indispensable amino acids proposed by the WHO are not reached, the amino acid score increased by the replacement of baker's flour by EVP. Hence, a high-fibre bread with elevated protein quality was achieved.

In addition, the structural differences could explain the differences in microbial shelf life. The protein matrix, putatively, also acts as a barrier for microbial attack on starch, prolonging the shelf life by two to three days. Apart from the structural influence, the increased water binding capacity of fibre and especially proteins restrict the accessibility to water and hinder the growth of mould [64].

5. Conclusions

BSG is a nutritious side-stream of the brewing industry that is used primarily for animal feed. Previously, using BSG as a food ingredient to enrich products, particularly cereal products, with highly nutritious dietary fibre and protein has been very challenging, more often than not leading to a significant decrease in final bread quality. The present study used two unique BSG-derived ingredients, EverVita FIBRA (EVF) and EverVita PRO (EVP), which differ in protein and dietary fibre content as well as particle size. These two ingredients performed very similarly in small inclusion levels but differently in bread systems with a 'high in fibre' addition level. The inclusion of EVF in high amounts resulted in outstanding bread characteristics such as high specific volume, soft crumb texture and a crumb structure comparable to a baker's flour control. While the impact of EVP on the techno-functional properties was less desired, it significantly increased the nutritional value of the bread by increasing protein content by 36% and fibre content by 3-fold, lowering pGI-values by up to 25%, elevating the amount of indispensable amino acids and hence the protein quality, and prolonging microbial shelf life. Notwithstanding this, correlation analysis revealed the influence of dough characteristics on bread quality which can help make controlled changes in the baking process or even the formulation for further optimisation using functional ingredients. In the past, several studies have addressed the impact of the most abundant dietary fibre in BSG, arabinoxylans, on bread quality and the interaction between fibre and gluten network. However, the impact of the protein fraction present in BSG has been neglected in the literature. This study provided a deep insight into the impact of several BSG-derived fractions, including protein, on bread quality. This work demonstrated the potential game-changing impact of brewing side-streams to sustainably enhance the nutritional value of food products.

Author Contributions: Conceptualization, E.K.A. and A.W.S.; formal analysis, A.W.S.; resources, S.M., G.C.; data curation, A.W.S.; writing—original draft preparation, A.W.S.; writing—review and editing, D.V., P.O., E.K.A.; visualization, J.J.A.; supervision, E.K.A.; project administration, D.V., S.M., G.C., P.O. All authors have read and agreed to the published version of the manuscript.

Funding: None.

Data Availability Statement: Not applicable.

Acknowledgments: The authors would like to thank Tom Hannon and Kate Hardiman for their technical support.

Conflicts of Interest: The authors declare no conflict of interest.

References

1. Ravindran, R.; Jaiswal, A.K. Exploitation of Food Industry Waste for High-Value Products. *Trends Biotechnol.* **2016**, *34*, 58–69. [CrossRef] [PubMed]
2. Fritsch, C.; Staebler, A.; Happel, A.; Márquez, M.A.C.; Aguiló-Aguayo, I.; Abadias, M.; Gallur, M.; Cigognini, I.M.; Montanari, A.; López, M.J.; et al. Processing, valorization and application of bio-waste derived compounds from potato, tomato, olive and cereals: A review. *Sustainability* **2017**, *9*, 1492. [CrossRef]
3. Stelick, A.; Sogari, G.; Rodolfi, M.; Dando, R.; Paciulli, M. Impact of sustainability and nutritional messaging on Italian consumers' purchase intent of cereal bars made with brewery spent grains. *J. Food Sci.* **2021**, *86*, 531–539. [CrossRef] [PubMed]
4. Schettino, R.; Verni, M.; Acin-albiac, M.; Vincentini, O.; Krona, A.; Knaapila, A.; Cagno, R.D.; Gobbetti, M.; Rizzello, C.G.; Coda, R. Bioprocessed Brewers ' Spent Grain Improves Nutritional and Antioxidant Properties of Pasta. *Antioxidants* **2021**, *10*, 742. [CrossRef]
5. Xiros, C.; Christakopoulos, P. Biotechnological potential of brewers spent grain and its recent applications. *Waste Biomass Valorization* **2012**, *3*, 213–232. [CrossRef]
6. Mussatto, S.I. Brewer's spent grain: A valuable feedstock for industrial applications. *J. Sci. Food Agric.* **2014**, *94*, 1264–1275. [CrossRef]
7. Steiner, J.; Procopio, S.; Becker, T. Brewer's spent grain: Source of value-added polysaccharides for the food industry in reference to the health claims. *Eur. Food Res. Technol.* **2015**, *241*, 303–315. [CrossRef]
8. Buffington, J. The Economic Potential of Brewer's Spent Grain (BSG) as a Biomass Feedstock. *Adv. Chem. Eng. Sci.* **2014**, *4*, 308–318. [CrossRef]

9. Lynch, K.M.; Steffen, E.J.; Arendt, E.K. Brewers' spent grain: A review with an emphasis on food and health. *J. Inst. Brew.* **2016**, *122*, 553–568. [CrossRef]
10. Mendis, M.; Simsek, S. Arabinoxylans and human health. *Food Hydrocoll.* **2014**, *42*, 239–243. [CrossRef]
11. Lu, Z.X.; Walker, K.Z.; Muir, J.G.; O'Dea, K. Arabinoxylan fibre improves metabolic control in people with type II diabetes. *Eur. J. Clin. Nutr.* **2004**, *58*, 621–628. [CrossRef] [PubMed]
12. Waters, D.M.; Jacob, F.; Titze, J.; Arendt, E.K.; Zannini, E. Fibre, protein and mineral fortification of wheat bread through milled and fermented brewer's spent grain enrichment. *Eur. Food Res. Technol.* **2012**, *235*, 767–778. [CrossRef]
13. Ktenioudaki, A.; Chaurin, V.; Reis, S.F.; Gallagher, E. Brewer's spent grain as a functional ingredient for breadsticks. *Int. J. Food Sci. Technol.* **2012**, *47*, 1765–1771. [CrossRef]
14. Stojceska, V.; Ainsworth, P.; Plunkett, A.; Ibanoğlu, S. The recycling of brewer's processing by-product into ready-to-eat snacks using extrusion technology. *J. Cereal Sci.* **2008**, *47*, 469–479. [CrossRef]
15. Guo, M.; Du, J.; Zhang, Z.; Zhang, K.; Jin, Y. Optimization of Brewer's spent grain-enriched biscuits processing formula. *J. Food Process Eng.* **2014**, *37*, 122–130. [CrossRef]
16. Amoriello, T.; Mellara, F.; Galli, V.; Amoriello, M.; Ciccoritti, R. Technological properties and consumer acceptability of bakery products enriched with brewers⇔ spent grains. *Foods* **2020**, *9*, 1492. [CrossRef]
17. Bannon, S.; Walton, J.; Flynn, A. The National Adult Nutrition Survey: Dietary fibre intake of Irish adults. *Proc. Nutr. Soc.* **2011**, *70*. [CrossRef]
18. European Food Safety Authority Scientific Opinion on Dietary Reference Values for carbohydrates and dietary fibre. *EFSA J.* **2010**, *8*, 1462. [CrossRef]
19. Fărcaş, A.C.; Socaci, S.A.; Tofană, M.; Mureşan, C.; Mudura, E.; Salanţă, L.; Scrob, S. Nutritional properties and volatile profile of brewer's spent grain supplemented bread. *Rom. Biotechnol. Lett.* **2014**, *19*, 9705–9714.
20. Ogunwale, S.L.; Otemuyiwa, I.O.; Ayodele, V.I.; Ilori, M.O.; Adewusi, S.R.A. Incorporation of spent grains in bread: Chemical and Nutritional properties. *J. Agroaliment. Process. Technol.* **2018**, *24*, 81–88.
21. Naibaho, J.; Korzeniowska, M. Brewers'spent grain in food systems: Processing and final products quality as a function of fiber modification treatment. *J. Food Sci.* **2021**, *86*, 1532–1551. [CrossRef] [PubMed]
22. Aprodu, I.; Simion, A.B.; Banu, I. Valorisation of the Brewers' Spent Grain through Sourdough Bread Making. *Int. J. Food Eng.* **2017**, *13*, 1–9. [CrossRef]
23. Sahin, A.W.; Hardiman, K.; Atzler, J.J.; Vogelsang-O'Dwyer, M.; Valdeperez, D.; Münch, S.; Cattaneo, G.; O'Riordan, P.; Arendt, E.K. Rejuvenated Brewer's Spent Grain: The impact of two BSG-derived ingredients on techno-functional and nutritional characteristics of fibre-enriched pasta. *Innov. Food Sci. Emerg. Technol.* **2021**, *68*, 102633. [CrossRef]
24. Hoehnel, A.; Bez, J.; Petersen, I.L.; Amarowicz, R.; Juśkiewicz, J.; Arendt, E.K.; Zannini, E. Enhancing the nutritional profile of regular wheat bread while maintaining technological quality and adequate sensory attributes. *Food Funct.* **2020**, *11*, 4732–4751. [CrossRef] [PubMed]
25. Ispiryan, L.; Heitmann, M.; Hoehnel, A.; Zannini, E.; Arendt, E.K. Optimization and Validation of an HPAEC-PAD Method for the Quantification of FODMAPs in Cereals and Cereal-Based Products. *J. Agric. Food Chem.* **2019**, *67*, 4384–4392. [CrossRef]
26. Regulation (EC) No 1924/2006. The European Parliament and of the Council on Nutrition and Health Claims Made on Food. 2006, pp. 9–25. Available online: https://eur-lex.europa.eu/legal-content/en/ALL/?uri=CELEX%3A32006R1924 (accessed on 14 April 2021).
27. Sahin, A.W.; Axel, C.; Zannini, E.; Arendt, E.K. Xylitol, mannitol and maltitol as potential sucrose replacers in burger buns. *Food Funct.* **2018**, *9*, 2201–2212. [CrossRef]
28. Dal Bello, F.; Clarke, C.I.; Ryan, L.A.M.; Ulmer, H.; Schober, T.J.; Ström, K.; Sjögren, J.; van Sinderen, D.; Schnürer, J.; Arendt, E.K. Improvement of the quality and shelf life of wheat bread by fermentation with the antifungal strain Lactobacillus plantarum FST 1.7. *J. Cereal Sci.* **2007**, *45*, 309–318. [CrossRef]
29. *Joint FAO/WHO/UNU Expert Consultation on Protein and Amino Acid Requirements in Human Nutrition: Report of a Joint FAO/WHO/UNU Expert Consultation*; World Health Organization: Geneva, Switzerland, 2007; Volume xi, 265p.
30. Brennan, C.S.; Tudorica, C.M. Evaluation of potential mechanisms by which dietary fibre additions reduce the predicted glycaemic index of fresh pastas. *Int. J. Food Sci. Technol.* **2008**, *43*, 2151–2162. [CrossRef]
31. Hoehnel, A.; Axel, C.; Bez, J.; Arendt, E.K.; Zannini, E. Comparative analysis of plant-based high-protein ingredients and their impact on quality of high-protein bread. *J. Cereal Sci.* **2019**, *89*. [CrossRef]
32. Hoehnel, A.; Bez, J.; Sahin, A.W.; Coffey, A.; Arendt, E.K.; Zannini, E. Leuconostoc citreum TR116 as a Microbial Cell Factory to Functionalise High-Protein Faba Bean Ingredients for Bakery Applications. *Foods* **2020**, *9*, 1706. [CrossRef]
33. Bouachra, S.; Begemann, J.; Aarab, L.; Hüsken, A. Prediction of bread wheat baking quality using an optimized GlutoPeak®-Test method. *J. Cereal Sci.* **2017**, *76*, 8–16. [CrossRef]
34. Hammann, F.; Schmid, M. Determination and quantification of molecular interactions in protein films: A review. *Materials* **2014**, *7*, 7975–7996. [CrossRef]
35. Nawrocka, A.; Szymańska-Chargot, M.; Miś, A.; Ptaszyńska, A.A.; Kowalski, R.; Waśko, P.; Gruszecki, W.I. Influence of dietary fibre on gluten proteins structure-A study on model flour with application of FT-Raman spectroscopy. *J. Raman Spectrosc.* **2015**, *46*, 309–316. [CrossRef]

36. Nawrocka, A.; Szymańska-Chargot, M.; Miś, A.; Wilczewska, A.Z.; Markiewicz, K.H. Effect of dietary fibre polysaccharides on structure and thermal properties of gluten proteins–A study on gluten dough with application of FT-Raman spectroscopy, TGA and DSC. *Food Hydrocoll.* **2017**, *69*, 410–421. [CrossRef]
37. Bock, J.E.; Damodaran, S. Bran-induced changes in water structure and gluten conformation in model gluten dough studied by Fourier transform infrared spectroscopy. *Food Hydrocoll.* **2013**, *31*, 146–155. [CrossRef]
38. Jones, R.; Erlander, S. Interaction between Wheat Proteins and Dextrans. *Cereal Chem.* **1967**, *44*, 447–456.
39. Gan, Z.; Gailliard, T.; Ellis, P.; Angold, R.; Vaughan, J. Effect of the outer bran layers on the loaf volume of wheat bread. *J. Cereal Sci.* **1992**, *15*, 151–163. [CrossRef]
40. Gómez, M.; Ronda, F.; Blanco, C.A.; Caballero, P.A.; Apesteguía, A. Effect of dietary fibre on dough rheology and bread quality. *Eur. Food Res. Technol.* **2003**, *216*, 51–56. [CrossRef]
41. Wang, M.; Van Vliet, T.; Hamer, R.J. How gluten properties are affected by pentosans. *J. Cereal Sci.* **2004**, *39*, 395–402. [CrossRef]
42. Wang, J.; Rosell, C.M.; Benedito de Barber, C. Effect of the addition of different fibres on wheat dough performance and bread quality. *Food Chem.* **2002**, *79*, 221–226. [CrossRef]
43. Koletta, P.; Irakli, M.; Papageorgiou, M.; Skendi, A. Physicochemical and technological properties of highly enriched wheat breads with wholegrain non wheat flours. *J. Cereal Sci.* **2014**, *60*, 561–568. [CrossRef]
44. Bordes, J.; Branlard, G.; Oury, F.X.; Charmet, G.; Balfourier, F. Agronomic characteristics, grain quality and flour rheology of 372 bread wheats in a worldwide core collection. *J. Cereal Sci.* **2008**, *48*, 569–579. [CrossRef]
45. Moore, W.E.; Carter, J.L. Protein-Carbohydrate Interactions at Elevated Temperatures. *J. Texture Stud.* **1974**, *5*, 77–88. [CrossRef]
46. Sivam, A.S.; Sun-Waterhouse, D.; Quek, S.Y.; Perera, C.O. Properties of bread dough with added fiber polysaccharides and phenolic antioxidants: A review. *J. Food Sci.* **2010**, *75*. [CrossRef] [PubMed]
47. Belton, P.S. On the elasticity of wheat gluten. *J. Cereal Sci.* **1999**, *29*, 103–107. [CrossRef]
48. Sluimer, P. Chapter 2: Basic ingredients. In *Principles of Breadmaking: Functionality of Raw Materials and Process Steps*; American Association of Cereal Chemists: St. Paul, MN, USA, 2005; pp. 17–47. ISBN 9781891127458.
49. Wang, M.; Hamer, R.J.; Van Vliet, T.; Gruppen, H.; Marseille, H.; Weegels, P.L. Effect of water unextractable solids on gluten formation and properties: Mechanistic considerations. *J. Cereal Sci.* **2003**, *37*, 55–64. [CrossRef]
50. Autio, K. Effects of cell wall components on the functionality of wheat gluten. *Biotechnol. Adv.* **2006**, *24*, 633–635. [CrossRef]
51. Silventoinen, P.; Rommi, K.; Holopainen-Mantila, U.; Poutanen, K.; Nordlund, E. Biochemical and Techno-Functional Properties of Protein- and Fibre-Rich Hybrid Ingredients Produced by Dry Fractionation from Rice Bran. *Food Bioprocess Technol.* **2019**, *12*, 1487–1499. [CrossRef]
52. Xiao, F.; Zhang, X.; Niu, M.; Xiang, X.; Chang, Y.; Zhao, Z.; Xiong, L.; Zhao, S.; Rong, J.; Tang, C.; et al. Gluten development and water distribution in bread dough influenced by bran components and glucose oxidase. *LWT* **2021**, *137*, 110427. [CrossRef]
53. Izydorczyk, M.S.; Hussain, A.; MacGregor, A.W. Effect of barley and barley components on rheological properties of wheat dough. *J. Cereal Sci.* **2001**, *34*, 251–260. [CrossRef]
54. Roels, S.P.; Cleemput, G.; Vandewalle, X.; Nys, M.; Delcour, J.A. Bread Volume Potential of Variable-Quality Flours with Constant Protein Level As Determined by Factors Governing Mixing Time and Baking Absorption Levels. *Cereal Chem.* **1993**, *70*, 319–323.
55. Collar, C.; Bollaín, C. Relationships between dough functional indicators during breadmaking steps in formulated samples. *Eur. Food Res. Technol.* **2005**, *220*, 372–379. [CrossRef]
56. Barak, S.; Mudgil, D.; Khatkar, B.S. Relationship of gliadin and glutenin proteins with dough rheology, flour pasting and bread making performance of wheat varieties. *LWT Food Sci. Technol.* **2013**, *51*, 211–217. [CrossRef]
57. Singh, J.; Dartois, A.; Kaur, L. Starch digestibility in food matrix: A review. *Trends Food Sci. Technol.* **2010**, *21*, 168–180. [CrossRef]
58. Oria, M.P.; Hamaker, B.R.; Schull, J.M. In vitro protein digestibility of developing and mature sorghum grain in relation to α-, β-, and γ-kafirin disulfide crosslinking. *J. Cereal Sci.* **1995**, *22*, 85–93. [CrossRef]
59. Hamaker, B.R.; Bugusu, B.A. Sorghum proteins and food quality. In Proceedings of the Afripro Conference, San Francisco, CA, USA, 25–29 August 2003; Volume 8. Available online: www.afripro.org.uk (accessed on 5 April 2021).
60. Rakhesh, N.; Fellows, C.M.; Sissons, M. Evaluation of the technological and sensory properties of durum wheat spaghetti enriched with different dietary fibres. *J. Sci. Food Agric.* **2015**, *95*, 2–11. [CrossRef]
61. Leinonen, I.; Iannetta, P.P.M.; Rees, R.M.; Russell, W.; Watson, C.; Barnes, A.P. Lysine Supply Is a Critical Factor in Achieving Sustainable Global Protein Economy. *Front. Sustain. Food Syst.* **2019**, *3*, 1–11. [CrossRef]
62. Friedman, M. Analysis, Nutrition, and Health Benefits of Tryptophan. *Int. J. Tryptophan Res.* **2018**, *11*. [CrossRef]
63. Shibata, K.; Shimada, H.; Kondo, T. Effects of feeding tryptophan-limiting diets on the conversion ratio of tryptophan to niacin in rats. *Biosci. Biotechnol. Biochem.* **1996**, *60*, 1660–1666. [CrossRef]
64. Mathlouthi, M. Water content, water activity, water structure and the stability of foodstuffs. *Food Control* **2001**, *12*, 409–417. [CrossRef]

 MDPI

Article

Extrusion Processing of Rapeseed Press Cake-Starch Blends: Effect of Starch Type and Treatment Temperature on Protein, Fiber and Starch Solubility

Anna Martin [1,*], Susanne Naumann [1], Raffael Osen [2], Heike Petra Karbstein [3] and M. Azad Emin [3]

[1] Department of Food Process Development, Fraunhofer Institute for Process Engineering and Packaging IVV, 85354 Freising, Germany; susanne.naumann@ivv.fraunhofer.de
[2] Singapore Institute of Food and Biotechnology Innovation, Agency for Science, Technology and Research (A*STAR), Singapore 138669, Singapore; Raffael_Osen@sifbi.a-star.edu.sg
[3] Institute of Process Engineering in Life Sciences, Chair of Food Process Engineering, Karlsruhe Institute of Technology, 76131 Karlsruhe, Germany; heike.karbstein@kit.edu (H.P.K.); azad.emin@kit.edu (M.A.E.)
* Correspondence: anna.martin@ivv.fraunhofer.de; Tel.: +49-8161-491-457

Abstract: For the valorization of oilseed press cakes into food products, extrusion can be used. A common way of applying the protein- and fiber-rich press cakes in directly expanded products is the combination thereof with starch, since starch gives a favourable texture, which correlates directly to expansion. To control product properties like expansion of protein and fiber-rich extruded products, the underlying physicochemical changes of proteins, fibers and starch due to thermomechanical input need to be comprehensively described. In this study, rapeseed press cake (RPC) was extruded and treated under defined thermomechanical conditions in a closed-cavity rheometer, pure and in combination with four starches. The impact of starch type (potato PS, waxy potato WPS, maize MS, high-amylose maize HAMS) and temperature (20/25, 80, 100, 120, 140 °C) on protein solubility, starch gelatinization (D_{gel}), starch hydrolysis (S_H) and fiber solubility of the blends was evaluated. The extrusion process conditions were significantly affected by the starch type. In the extruded blends, the starch type had a significant impact on the protein solubility which decreased with increasing barrel temperature. Increasing barrel temperatures significantly increased the amount of soluble fiber fractions in the blends. At defined thermomechanical conditions, the starch type showed no significant impact on the protein solubility of the blends. Therefore, the observed effects of starch type on the protein solubility of extruded blends could be attributed to the indistinct process conditions due to differences in the rheological properties of the starches rather than to molecular interactions of the starches with the rapeseed proteins in the blends.

Keywords: canola; protein solubility; dietary fiber; starch gelatinization; extrusion; expansion; biopolymers; closed-cavity rheometer

Citation: Martin, A.; Naumann, S.; Osen, R.; Karbstein, H.P.; Emin, M.A. Extrusion Processing of Rapeseed Press Cake-Starch Blends: Effect of Starch Type and Treatment Temperature on Protein, Fiber and Starch Solubility. *Foods* **2021**, *10*, 1160. https://doi.org/10.3390/foods10061160

Academic Editor: Zhengyu Jin

Received: 8 April 2021
Accepted: 18 May 2021
Published: 21 May 2021

Publisher's Note: MDPI stays neutral with regard to jurisdictional claims in published maps and institutional affiliations.

1. Introduction

The valorization of food production by-products into directly expanded food products using extrusion has been a well-established and well-studied technology in the past years [1–9]. In response to the consumer demand for healthy and sustainable products, protein- and fiber-rich by-products of the food production chain have been utilized in order to aim at beneficial nutritional profiles, as well as designated texture properties [10]. A promising source of protein and fiber is rapeseed press cake (RPC), which constitutes the residue after the oil-pressing of rapeseeds. Due to its high availability and some nutritional limitations that restrict its inclusion level, RPC is relatively inexpensive compared to other press cakes like soybean or sunflower. RPC has been applied as an ingredient in blends in order to investigate its impact on the extruder response, rheological properties, and physical quality properties of extruded food and feed [11–15]. However, its maximum level of incorporation, especially in extruded food products, is restricted due to two main factors—the presence of antinutritional components (ANFs) and its effect on product quality, such as texture properties. To overcome the hurdle of large quantities of ANFs in RPC, a number of studies have investigated the impact of different treatments on ANF reduction and have successfully accomplished a significant reduction of glucosinolates, tannins, phytic acids and raw fiber [16–20]. It can therefore be assumed that in the next few years, the use of RPC with reduced ANF levels will allow the production of RPC-based extruded products for human consumption. The second major restriction of using RPC in extruded products is the limited knowledge of its effect on product quality, particularly on the product texture that is defined by its expansion properties. Expansion is known to be driven by the sudden exceedance of water vapor pressure at the extruder die exit. Expansion properties highly depend on the extrusion process parameters, like barrel temperature, moisture content or screw speed, that in turn influence the rheological, chemical and physicochemical properties of the extruded matrix.

High expansion properties of extruded snacks are mostly generated by extrusion processing of starch-based formulations, such as potato, rice, wheat or maize starch [21]. When RPC (10–70 g/100 g) was added to starch blends in previous studies, severe changes in expansion properties, as well as in rheological and physicochemical properties were monitored. In our recent study [15] we reported that up to 70 g/100 g RPC could be implemented in potato starch blends, resulting in a high degree of expansion. However, a combination of maize starch and RPC (10–40 g/100 g) resulted in a lower degree of expansion, and the application of RPC resulted in changes of rheological and physicochemical properties of the blends and the corresponding extruded samples, like viscous and elastic properties, water absorbance or water solubility [14]. In these studies, the underlying physicochemical transformations of RPC and starch have not been described yet, but can be related to a number of heat- and shear-induced reactions that take place due to thermomechanical treatment.

RPC exhibits protein content between 19–40 g/100 g [22–24]. In general, plant protein fortification of starch-based extruded products has been reported to result in significant changes of the final product properties due to unfolding, realigning, hydrolysation, denaturation and cross-linking of the proteins with each other or with other ingredients, like starch, sugar or dextrin molecules [25,26]. Denaturation and aggregation of plant proteins induced by thermomechanical input has been reported to reduce the protein solubility in water or salt buffers [27]. In particular, Zhang et al. [28] reported that extrusion-processing of rapeseed protein meal resulted in significant protein aggregation, and consequently in a lower amount of extractable protein compared to untreated samples. Matthey and Hanna [29] proposed that protein–starch interactions can in turn inhibit the degradation of starch during extrusion, especially because proteins adjust the water distribution in the melt. Starch degradation, indicated by the degree of gelatinization, is known to be affected by the botanical origin, as well as by the amylose/amylopectin ratio of the starch type [30]. Some studies suggest that amylopectin and amylose may physically interact or form new bonds with, for example, proteins during thermomechanical treatment, wherefore product properties like expansion are influenced [31]. Therefore, it can be expected that with the combination of RPC and starch in blends, reactions of rapeseed protein and the competition for water during thermomechanical treatment between starch and RPC will affect the degree of starch degradation and the solubility of rapeseed proteins.

Besides proteins, RPC exhibits about 36 g/100 g of total dietary fiber content, whereof 88 g/100 g are soluble and 12 g/100 g are insoluble [15]. A transformation of rapeseed fiber components from insoluble to soluble can be expected, as reported in previous studies [32–34]. This would in turn have an effect on the rheological properties, and consequently on the final product properties of extruded RPC/starch products.

In order to increase the amount of RPC in starch-based extruded snacks and to control the quality of the extruded products, the underlying physicochemical changes of these biopolymers need to be investigated in relation to the applied process conditions.

Therefore, in this study, the impact of extrusion processing on the physicochemical properties of rapeseed protein, rapeseed fiber, as well as of starch was evaluated. RPC as a pure component and combined in blends with four starch types was investigated, both untreated and after thermomechanical treatment.

It was expected that the extruder response (SME and product temperature) during extrusion would be affected by the blend composition of the starch type and the resulting rheological properties, respectively, although the process conditions (barrel temperature, mass flow rate, screw speed) were kept constant. Therefore, to overcome the hurdle of indistinct temperatures and shear rates applied to the blends in the extruder, a closed-cavity rheometer was used in order to execute thermomechanical treatments at defined temperatures and shear rates.

2. Materials and Methods

2.1. Material and Preparation of Blends

Cold-pressed 00-type RPC was kindly provided by Teutoburger Ölmühle (Ibbenbüren, Germany). The temperature during pressing did not exceed 60 °C. The pH of RPC was 5.97 ± 0.25. Potato starch (PS) and waxy potato starch (WPS Eliane^{TM} 100) were kindly provided by Avebe (Veendam, Netherlands). Maize starch (MS) and high-amylose maize starch (HAMS) (Hylon^{TM} VII PCR) were kindly provided by Ingredion (Hamburg, Germany). The amylose content of the starches, as reported in the specifications of the suppliers, were 25, 1, 26 and 67 g/100 g for PS, WPS, MS and HAMS. The RPC was milled to <500 μm before it was mixed in a 70:30 ratio wet basis (w.b.) with PS, WPS, MS or HAMS in a Spiral-Mixer SP 12 (DIOSNA Dierks & Söhne GmbH, Osnabrück, Germany) for 60 min, followed by an incubation at 20 °C for at least 8 h. Prior to extrusion, the dry matter content of the mixtures was analyzed (MA 40, Sartorius AG, Göttingen, Germany) as described in the German Food Act [35].

The moisture content of the materials and extrudates was determined according to the German Food Act [35]. The protein content was analyzed based on the Dumas method according to the German Food Act [35] using a TruMac N Protein Analyzer (LECO, St. Joseph, MI, USA). The ash content was determined according to the AOAC International method 945.46 [36]. The crude fiber content was determined according to the AOAC International method 962.09 [37]. The starch content was determined as previously described [38]. Water absorption (g/g) of the raw materials and extruded samples was analyzed according to the AACC method 56–20.01, and water solubility (%) was determined as previously described [39]. The particle sizes of MS, RPC and RP were determined using a Malvern Mastersizer S Long Bed Version 2.15 laser diffraction particle size analyzer (Malvern Instruments, Malvern, UK) as previously described [40]. Analyses were carried out in duplicate.

2.2. Extrusion Processing

The extrusion process was carried out using a 26 mm pilot scale twin screw co-rotating extruder with a L/D ratio of 25/1 (ZSK 26 Mc, Coperion, Stuttgart, Germany). The mass flow rate was kept at 10 kg/h, the moisture content of the melt was set to a 29 g/100 g dry matter basis (d.m.), the screws rotated with 300 rpm, and the temperature of the last barrel segment was set to 20, 80, 100, 120 or 140 °C. A detailed temperature and screw profile can be found in Table 1. After extrusion, the samples were dried in an oven (Thermo Scientific Heraeus UT 6760, Thermo Electron LED GmbH, Langenselbold, Germany) at 40 °C for 24 h, milled at 14.000 rpm to <500 μm (ZM 200, Retsch, Haan, Germany), vacuum-sealed and stored at 20 °C until further analyses.

2.3. Thermomechanical Treatment at Defined Temperatures and Shear Rates

Defined thermomechanical conditions were applied using a closed-cavity rheometer (RPA elite, TA instruments, New Castle, DE, USA). The cavity can be pressurized (4.5 MPa) and sealed and the device allows the analyses of low moisture samples at high temperatures without water vaporization or material slippage [41]. Before analyses, the samples were brought to 29 g/100 g d.m. moisture content by mixing them with deionized water in a Thermomix (Vorwerk, Wuppertal, Germany). Afterwards, the samples were incubated in a fridge (4 °C) for at least 24 h to ensure homogeneous water distribution. For rheological analyses, 6 g of each sample were brought to room temperature and placed on the cone.

Table 1. Experimental set-up for extrusion-processing and treatments at defined thermomechanical conditions.

Extrusion Processing	Barrel Temperature T_B of Barrel Segments 2–6 (°C)					Residence Time (s)	Shear Rate $\dot{\gamma}$ (s^{-1})
1	20	20	20	20	20	Approx. 60	15–2430 *
2	60	80	80	80	80	Approx. 60	15–2430 *
3	60	80	100	100	100	Approx. 60	15–2430 *
4	60	80	100	120	120	Approx. 60	15–2430 *
5	60	100	120	140	140	Approx. 60	15–2430 *
Defined Thermomechanical Treatment	**Treatment Temperature T_T (°C)**					**Treatment Time T_t (s)**	**Shear Rate $\dot{\gamma}$ (s^{-1})**
1	25					60	50
2	80					60	50
3	100					60	50
4	120					60	50
5	140					60	50

* Values were investigated by Emin and Schuchmann [42] using CFD simulation based on an extruder of the same type, a similar screw configuration, and the same screw speed as used in this study. Shear rates differ largely as a function of location of the melt in the different screw sections. Therefore, values are only given as an estimation.

Extrusion-like conditions were simulated by applying temperatures corresponding to the barrel temperatures during extrusion. Isothermal time sweep tests at 25, 80, 100, 120 or 140 °C were performed at a shear rate of 50 s^{-1} (corresponding to f = 10 Hz and γ = 80%, non-LVE region). The treatment time was 60 s to mimic the residence time in the used extruder. Since the temperature in the measuring chamber of the closed-cavity rheometer is cooled by air, the lowest possible treatment temperature was 25 °C. For each treatment, at least five samples were collected. After treatment, the samples were treated as the extruded samples (see Section 2.2) and investigated in regard of the extractable protein content and protein solubility (see Section 2.4.1).

2.4. *Physicochemical Properties*

2.4.1. Extractable Protein Content and Protein Solubility

The content of the salt-soluble protein fraction in the untreated and thermomechanically treated samples in NaCl was determined at pH 4, 7 and 11 and is further referred to as extractable protein content. Generally, salt buffers have been used in previous research in order to extract protein in its native state [43].

1500, 2000 or 3000 mg of each sample were placed in 50 mL beakers. NaCl of 0.1 M was added, and the pH was adjusted as desired with NaOH for alkaline samples and HCl for acidic samples. The samples were placed in a magnetic stirring plate at 200 rpm for 1 h. After 30 min and after 1 h, the pH was checked and adjusted if necessary. Each sample was placed in a 50 mL volumetric flask and filled up to the mark with 0.1 M NaCl.

After the protein extraction step, 20 mL of the solution were placed in centrifuge tubes and centrifuged for 15 min at 15 °C and 20.000 rpm. The supernatant was filtered (Whatman^{TM}, diameter 150 mm, pore size 4–12 µm) and stored at −20 °C until analysis.

The Dumas method established by Eblinger et al. [44] was employed to determine the protein content in the supernatants based on its dry matter. Analyses were carried out in triplicate.

The protein solubility was calculated using Equation (1). V represents the initial volume of the sample (50 mL), PC refers to the protein content present in the supernatant as determined by the Dumas method, m_s refers to the initial mass of the sample, DM_s is the dry matter of the sample, and PC_{dm} is the protein content of the dry matter of the sample.

$$Psol = (V \times PCsupernatant \times 100)/(ms \times DMs \times PCdm) \quad (1)$$

2.4.2. Starch Gelatinization

The gelatinization temperature T_{gel} and enthalpy Δ H of RPC/starch blends were analyzed by using a differential scanning calorimeter (DSC) (Q2000, TA Instruments, New Castle, DE, USA). Slurries were prepared with the unextruded and extruded samples to obtain a final moisture content of 60 g/100 g (w.b.) using a Thermomix (Vorwerk, Wuppertal, Germany). Slurry amounting to 15–20 g were placed into aluminum pans and sealed for analysis. A sealed empty pan was used as a reference. A heating ramp of 20–120 °C with a heating rate of 10 °C/min was applied for one scan. Analysis was carried out using the TA Universal Analysis software (4.4 A, TA Instruments, New Castle, DE, USA). T_{gel} was taken as the peak temperature of the gelatinization endotherm. The degree of gelatinization D_{gel} was calculated according to Equation (2), where Δ H_0 refers to the gelatinization enthalpy of the unextruded formulation and ΔH_{gel} indicates the gelatinization enthalpy of the extruded formulations. Analysis was carried out in triplicate.

$$D_{gel} = (\Delta H_0 - \Delta H_{gel})/\Delta H_0 \times 100 \quad (2)$$

2.4.3. Starch Hydrolysis

The total hydrolyzed starch of RPC/starch blends was analyzed by using a Starch UV Test kit (r-biopharm, Darmstadt, Germany) according to Beutler et al. [38]. Starch was hydrolyzed using amyloglucosidase, hexokinase and glucose-6-phosphate dehydrogenase. Analysis was carried out at least in duplicate for unextruded samples and samples extruded at 100 °C or 140 °C.

2.4.4. Soluble and Insoluble Dietary Fiber Analysis

Soluble and insoluble fiber contents of non-extruded RPC and RPC/starch blends were analyzed by enzymatic-gravimetric analysis according to AOAC 991.43. Analyses were carried out in triplicate.

2.4.5. Statistical Analysis

Data were analyzed using OriginPro (2018 b, OriginLab Corporation, Northampton, MA, USA) by means of two-way analysis of variance (ANOVA). Statistics were considered significant when $p \leq 0.05$. When appropriate, means were compared using Tukey's honest significance test.

3. Results and Discussion

3.1. Chemical Analysis

Table 2 reports the chemical composition of RPC and RPC/starch blends. The RPC/starch blends exhibited very similar protein, ash, starch and lipid contents. TDF content increased in the order WPS30/RPC70, PS30/RPC70, MS30/RPC70 and HAMS30/RPC70. MS blends exhibited the highest, and PS blends the lowest amount of SDF. HAMS blends had a significantly higher amount of IDF compared to blends with PS, MS und WPS that showed the least amount of IDF.

Table 2. Chemical composition of rapeseed press cake (RPC100) and blends of 30 g/100 g potato starch (PS), waxy potato starch (WPS), maize starch (MS) and high-amylose maize starch (HAMS) in combination with 70 g/100 g RPC.

Chemical Composition	RPC100	PS30/RPC70	WPS30/RPC70	MS30/RPC70	HAMS30/RPC70
Dry matter (d.m.) (%)	95.10 ± 0.03 [a]	92.18 ± 0.01 [b]	91.96 ± 0.08 [b]	93.50 ± 0.06 [c]	93.41 ± 0.03 [c]
Protein (% d.m.)	38.20 ± 0.30 [a]	27.40 ± 0.11 [b]	27.38 ± 0.29 [b]	26.99 ± 0.038 [b]	27.28 ± 0.029 [b]
Ash (% d.m.)	7.30 ± 0.02 [a]	4.72 ± 0.02 [b]	4.79 ± 0.01 [b]	4.67 ± 0.025 [b]	4.41 ± 0.55 [b]
TDF (% d.m.)	35.67 ± 5.19 [a]	10.28 ± 2.82 [b]	9.01 ± 1.11 [b]	11.51 ± 0.14 [b]	16.92 ± 0.37 [c]
SDF (% d.m.)	4.17 ± 1.55 [a]	1.82 ± 0.13 [b]	1.95 ± 0.05 [b]	3.38 ± 0.11 [c]	2.97 ± 0.28 [c]
IDF (% d.m.)	31.51 ± 4.96 [a]	8.47 ± 2.81 [b]	7.06 ± 1.11 [b]	8.14 ± 0.07 [b]	13.95 ± 0.24 [c]
Starch (% d.m.)	3.00 ± 0.02 [a]	29.66 ± 0.014 [b]	28.93 ± 0.052 [b]	29.39 ± 0.021 [b]	27.26 ± 0.038 [b]
Lipid (% d.m.)	23.40 ± 0.90 [a]	16.65 ± 0.25 [b]	16.16 ± 0.02 [b]	15.71 ± 0.03 [b]	16.17 ± 0.11 [b]
Particle size distribution ($Dv_{0.5}$/µm)	261.1 ± 4.5 [a]	87.0 ± 1.4 [b]	160.7 ± 13.1 [c]	131.5 ± 6.7 [d]	141.1 ± 6.8 c [e]

Mean values with different superscript letters within one row indicate significant differences ($p < 0.05$) based on a one-way analysis of variance (ANOVA). Where appropriate, the mean values were compared using Tukey's honest significance test.

The effect of starch type on the amounts of dietary fiber in untreated samples is likely to be an effect of the analysis method and the presence of resistant starch. Dietary fiber analysis in this study was initiated with an enzymatic starch digestion step (initiated by amylase) at the beginning of the extraction, and resistant starches, if present, remain unaffected by this treatment. It is known from previous studies that uncooked high-amylose starches are more resistant to enzymatic hydrolysis than high-amylopectin starches [45].

3.2. Extruder Response

Table 3 reports the extruder response as a function of barrel temperature and starch type generated by the RPC/starch blends. The addition of starch to RPC had a significant effect on the SME and product temperature. At a T_B of 20 °C, the product temperature increased with starch addition irrespective of starch type. At a T_B of 80 °C and 120 °C, RPC generated a similar product temperature as HAMS30/RPC70 and WPS30/RPC70. RPC as a single component could not be extruded at a T_B of 140 °C due to severe extruder clogging. We assume that this was due to the high water absorption of fiber components in RPC.

Table 3. Specific mechanical energy (SME) input and product temperature of extruded rapeseed press cake (RPC100) and blends of 30 g/100 g potato starch (PS), waxy potato starch (WPS), maize starch (MS) or high-amylose maize starch (HAMS) in combination with 70 g/100 g RPC as a function of barrel temperature T_B.

			SME (Wh/kg)		
T_B (°C) in segment 6	RPC100	PS30/RPC70	WPS30/RPC70	MS30/RPC70	HAMS30/RPC70
20	61.23 ± 1.5 [a]	98.13 ± 6.16 [b]	104.32 ± 2.95 [b]	60.68 ± 0.63 [a]	69.16 ± 0.63 [a]
80	38.08 ± 1.2 [a]	76.73 ± 0.92 [b]	83.70 ± 1.45 [b]	47.76 ± 4.02 [c]	53.22 ± 0.53 [d]
100	37.37 ± 1.1 [a]	74.89 ± 1.29 [b]	78.52 ± 0.87 [b]	45.85 ± 2.40 [c]	46.25 ± 0.44 [c]
120	36.57 ± 1.0 [a]	72.486 ± 1.53 [b]	74.82 ± 1.16 [b]	45.42 ± 2.73 [c]	40.74 ± 0.27 [ac]
140	-	65.24 ± 2.25 [a]	68.94 ± 1.42 [a]	36.52 ± 1.10 [b]	38.40 ± 0.50 [b]
			Product Temperature (°C)		
T_B (°C) in segment 6	RPC100	PS30/RPC70	WPS30/RPC70	MS30/RPC70	HAMS30/RPC70
20	30.0 ± 0.5 [a]	45.1 ± 0.5 [b]	52.2 ± 0.5 [ab]	43.7 ± 1.0 [c]	42.0 ± 0.3 [c]
80	72.1 ± 0.4 [a]	85.5 ± 0.5 [b]	90.2 ± 0.4 [c]	80.8 ± 0.8 [d]	72.0 ± 0.7 [a]
100	88.1 ± 0.5 [a]	104.0 ± 0.3 [b]	100.0 ± 0.2 [b]	98.0 ± 0.6 [b]	90.3 ± 1.2 [a]
120	108.2 ± 0.4 [a]	116.1 ± 0.1 [b]	108.8 ± 0.5 [a]	111.3 ± 0.3 [c]	100.0 ± 1.0 [d]
140	-	135.3 ± 0.2 [a]	123.4 ± 0.4 [b]	124.3 ± 0.2 [b]	112.3 ± 1.2 [c]

Mean values with different superscript letters within one row indicate significant differences ($p < 0.05$) based on a one-way analysis of variance (ANOVA). Where appropriate, the mean values were compared using Tukey's honest significance test.

A comparison of all RPC/starch blends exhibited that starch type and T_B had a significant impact on SME ($p < 0.01$). At all T_B, RPC/WPS blends exhibited the highest SME values, and RPC/MS blends the lowest ones.

RPC/PS and RPC/WPS blends exhibited higher SME at all T_B than RPC/MS and RPC/HAMS blends. With increasing T_B, the SME significantly decreased in all RPC/starch blends, which was due to a temperature-induced decrease of melt viscosity. PS and WPS gelatinized at a lower temperature compared to maize starches (see Section 3.5). This may have resulted in the release of amylose and amylopectin molecules in PS/WPS at a lower T_B compared to MS and HAMS increasing the SME.

Furthermore, the crystalline regions of the starches can act as "rigid structures" during extrusion, resulting in a more pronounced friction in the melt or between the melt and the extruder barrel [46]. Since WPS blends exhibit the highest amylopectin content and therefore a higher number of large molecules and a greater surface area compared to other starches, a higher viscosity and therefore SME compared to low-amylopectin blends like HAMS30/RPC70 is expected. This observation can be linked to Section 3.5, where WPS blends exhibited the highest gelatinization enthalpy, an indication for the largest amount of transformed amylopectin molecules.

3.3. *Impact of Thermomechanical Treatment on Protein Solubility of RPC*

3.3.1. Extruded RPC

Temperature and shear during thermomechanical treatment can lead to the denaturation of proteins and to the formation of protein-linkages, which affect the extractability and solubility of proteins. With elevated temperatures, native proteins unfold and expose new reactive binding sites, that aggregate and form bonds with other proteins, polysaccharides, lipids, fiber or secondary plant metabolites [43].

The extractable protein content based on the dry matter of the extract can be used as an indirect measure for the content of other solubilized macronutrients besides the proteins present in the extracts. If, for example, starch solubilizes in NaCl at the given conditions, it will contribute to the dry matter of the extract and lead to a lower amount of extractable protein content. Moreover, the protein solubility reports the amount of solubilized protein relative to the absolute protein content in the sample before extraction.

Figure 1a,b shows the content of extractable protein and the protein solubility of extruded RPC at pH 4, 7 and 11 as a function of T_B. Overall, the amount of extractable protein and the protein solubility from untreated RPC were low at an acidic pH, and were highest at strong alkaline conditions. The highest amount of extractable protein and solubility among all samples was found for untreated RPC at pH 11.

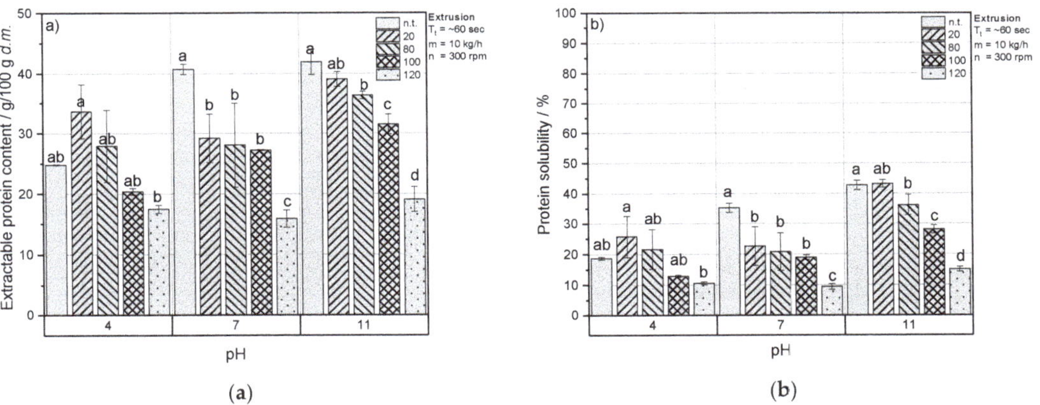

Figure 1. (**a**) Extractable protein content (g/100 g d.m.) and (**b**) protein solubility (%) at a pH of 4, 7 or 11 as a function of barrel temperature T_B of extruded rapeseed press cake (RPC). Mean values with different superscript letters comparing the effect of barrel temperature T_B within pH 4, 7 or 11 indicate significant differences ($p < 0.05$) based on a two-way analysis of variance (ANOVA). Where appropriate, the mean values were compared using Tukey's honest significance test. n.t. = not treated.

At a pH of 4, a T_B of 20 and 80 °C increased the content of extractable protein and solubility in RPC compared to the untreated RPC.

When T_B was set from 80 to 100 or 120 °C, the amount of extractable protein and the protein solubility decreased with increasing T_B. At a pH of 7 and 11, an increasing T_B resulted in a decreased amount of extractable protein content and solubility. At pH 7, the extractable protein content and solubility for RPC extruded at 20 °C was significantly lower as for untreated RPC, and a significant decrease was observed when T_B increased from 100 to 120 °C. At pH 11, with an increase of T_B to 20, 80 and 100 °C, the amount of extractable protein content decreased to a similar level at each step of temperature increase; however, a significant decrease in extractable protein content was monitored when T_B increased from 100 to 120 °C.

Regardless of T_B, the detectable and solubilized protein content in RPC extracts was highest for samples analyzed at pH 11.

The lower extractable protein content and solubility of untreated RPC at pH 4 compared to pH 7 and 11 is in agreement with previous research [22,47,48]. Fetzer et al. [22] observed that for cold-pressed rapeseed press cake, minimum protein solubility was observed in the acidic pH range, while the protein solubility was higher in the alkaline pH range. This can be attributed to the structural properties of the two major storage proteins in rapeseed.

The two most dominant proteins found in Brassica napus (rapeseed) are the 11S globulin cruciferin (300 kDa [49]) and the 1.7–2S albumin napin (12.5–14.5 kDa [49]), of which cruciferin constitutes 50–60% and napin 20–40% of the total protein accumulated in rapeseed. The tertiary structure of cruciferin is very pH-unstable (PI 7.2), even at ambient temperature, and unfolds at a low pH, whereas napin is stable in a wider pH range (PI 11), because of its helical secondary structure [50]. At acidic conditions (pH 4), only napins (10–16 kDa) were reported in RPC protein extracts in previous studies with minor fractions of cruciferins ($\geq$18 kDa) [22].

Since cruciferin constitutes the major protein fraction in rapeseed, consequently, the amount of extractable protein at pH 4 is lower compared to pH 7 and 11, where cruciferin predominantly occurs in protein extracts. At high-alkaline conditions, protein hydrolysis (proteolysis) can appear, where cruciferin bands can be degraded, while napin bands are only slightly affected, which in turn increases the overall protein solubility in this pH range. The ratio of soluble cruciferin and napin can vary at neutral pH (2.1:1 to 2.6:1), wherefore pH 7 gives the most mediocre picture for the amount of extractable protein. [22].

A reduced solubilisation of proteins due to extrusion processing was described in a number of studies and was attributed to protein cross-linking induced by thermal denaturation and Maillard reactions [43,51]. Napin was shown to be more stable against denaturation in a wider temperature range than cruciferin due to some structural features [49], while cruciferin denatures at lower temperatures, indicating that the contribution of unfolding, aggregation and cross-linking of the major rapeseed storage protein cruciferin on the overall reduced protein solubility is high. Cruciferin and napin of rapeseed protein isolate showed a denaturation temperature of 84 and 102 °C in previous research [50]. It can therefore be assumed that at a T_B of 100 °C, cruciferin is already denatured, whereas napin might still be native or partially native. When T_B increased from 100 to 120 °C, it is likely that both main proteins were denatured and might have formed new protein bonds, which explains the significant decrease of extractable protein content and solubility at pH 7 and 11. The significant decrease in extractable protein at pH 4 with an increase of T_B from 80 to 100 °C supports these assumptions, since napin, which denatures at temperatures > 80 °C, contributes majorly to the protein fractions extractable at acidic pH.

Additionally to protein linkages that are formed due to protein unfolding and cross-linking, Maillard reactions products (e.g., (methy-) glyoxal) can induce protein cross-linking through reactions with lysyl, agrinyl or tryptophanyl residues. We observed that the RPC samples changed from a light yellow colour to brown due to thermomechanical treatment, which strengthens this hypothesis.

The amount of extractable RPC proteins at pH 7 and 11 (Figure 1a) was slightly higher as the protein content was detected in the dry RPC powder (see Table 2). This indicates that during extraction, RPC proteins dissolved well in NaCl, and NaCl-insoluble components of RPC (e.g., insoluble dietary fiber) accumulated as sediments during extract centrifugation, wherefore the dry matter composition of the extract shifted to a more protein-rich composition.

Overall, the protein solubility of RPC was lower compared to rapeseed meal investigated in previous studies [52,53]. Fetzer et al. [22] reported a protein solubility of 35, 55 and 65% for cold-pressed rapeseed meal at pH 4, 7 and 11. This might be due to the higher fat content of the RPC used in our study (23.4 g/100 g) compared to the study of Fetzer et al. (2.8 g/100 g), thus enhancing complex formation of lipids what can hinder the solubilisation of proteins [54]. Another reason can be the presence of phytic acid in the

RPC used in our study. Phytic acid accumulates in large quantities in RPC and can form complexes with proteins which decrease the protein solubility [55,56].

3.3.2. Thermomechanical Treatment of Rpc under Defined Conditions

Figure 2a,b illustrate the amount of extractable protein and protein solubility (pH 7) of RPC treated at defined thermomechanical conditions.

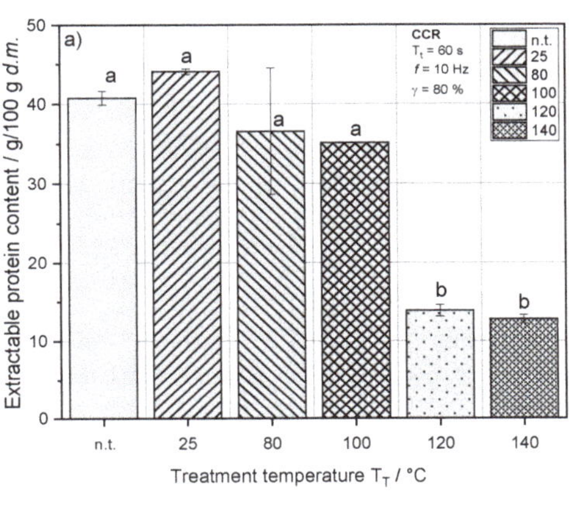

(**a**)

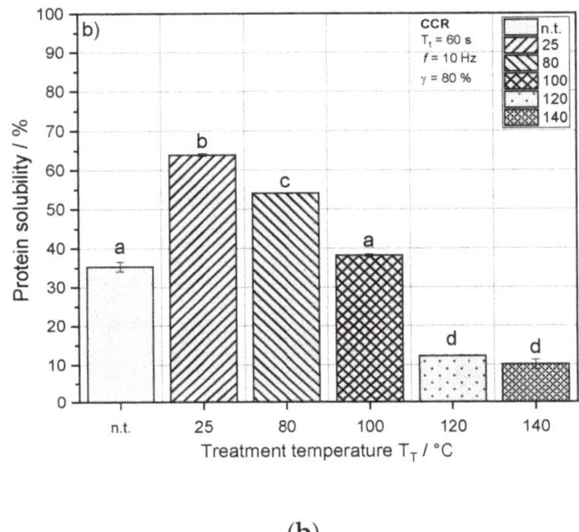

(**b**)

Figure 2. (**a**) Extractable protein content (g/100 g d.m.) and (**b**) protein solubility (%) at a pH of 7 as a function of the treatment temperature T_T of rapeseed press cake (RPC). Mean values with different superscript letters indicate significant differences ($p < 0.05$) based on a one-way analysis of variance (ANOVA). Where appropriate, the mean values were compared using Tukey's honest significance test. n.t. = not treated.

RPC exhibited a slightly higher amount of extractable protein and solubility when treated at 25 °C compared to the untreated sample. This effect was also seen for extruded RPC at a T_B of 20 °C and extracted at pH 4. At a T_T and T_B of 25 and 20 °C, respectively, the effect of the mechanical energy input on protein extractability of RPC proteins dominates over the impact of temperature (see extruder response reported in Table 3 for the sample RPC100). An increase of extractable protein and solubility due to mechanical input can be explained by the disruptive impact of shear on the cellular structure of RPC, wherefore the solvent during protein extraction can access a greater surface area of rapeseed components resulting in an increased content of dissolved protein in the extract.

With T_T > 25 °C, the proteins were less extractable and protein solubility decreased, which can be explained by the onset of protein aggregation reactions at these temperatures and a higher mobility of macromolecules due to elevated temperatures, increasing the chance for the formation of new linkages [57]. The temperature during the pressing of rapeseed oil, where RPC is generated, is continuously below 60 °C. Considering that the denaturation temperatures of the main rapeseed protein fractions napin and cruciferin are higher than 60 °C, a relatively high protein nativity of the rapeseed proteins in RPC can be assumed. Martin et al. [15] reported reaction onset temperatures of rapeseed components to be >70 °C, indicated by an increase of the complex modulus (G*) during a temperature sweep measurement in a closed-cavity rheometer. The experimental set-up and material conditions in this study were set alike the conditions in the present study, wherefore it can be assumed that with an increase of T_T from 25 to 80 °C, aggregation reactions of rapeseed proteins are induced, and consequently, the extractability of proteins is reduced.

At a T_T of 80 and 100 °C, similar amounts of protein could be extracted; however, protein solubility decreased significantly. Furthermore, significantly less protein was extracted and solubilized in RPC treated at 120 and 140 °C. This can again be linked to the high denaturation temperature of napin that is only exceeded at a T_T of 120 °C. Extruded samples exhibited a very similar rapid decrease of extractable protein content and solubility at a temperature of 120 °C (see Figure 1a,b).

An increase of T_T from 120 to 140 °C only slightly decreased the amount of extractable proteins and the protein solubility in RPC. It can be assumed that with the denaturation of napin, the major part of possible formations of new protein linkages is achieved. The slight decrease of extractable protein content at T_T of 140 °C may be attributed to reactions of other minor proteins in RPC, that are oleosins, lipid transfer proteins and protease inhibitors [58].

3.3.3. Comparison of Extrusion-Processing and Defined Thermomechanical Conditions

Compared to extruded RPC, samples treated at defined thermomechanical conditions exhibited significantly higher protein solubility up to a treatment temperature of 100 °C. The most significant decrease of protein solubility was observed when T_B or T_T were increased from 100 to 120 °C.

At 120 °C, the protein solubility of RPC was slightly lower in extruded samples compared to samples treated at defined thermomechanical conditions. A lower protein solubility of extruded RPC compared to RPC treated under defined conditions can be attributed to the higher shear forces in the extruder barrel, leading to higher local temperatures, enhancing protein aggregation reactions. It can be assumed that at temperatures above 100 °C, thermal energy input is dominating over the impact of shear stress and causing a severe decrease of protein solubility, again correlating with the denaturation temperatures of rapeseed proteins.

3.4. Impact of Thermomechanical Treatment and Starch Addition on Protein Solubility of RPC

3.4.1. Extractable Protein Content and Solubility of RPC/Starch Blends at Neutral pH

Figure 3 shows the extractable protein content and solubility of RPC/starch blends as a function of treatment temperature at a defined shear rate of 50 s^{-1}.

Untreated Blends

The amount of extractable proteins from untreated RPC/starch blends was significantly influenced by the starch type (Figure 3a,b), and so was the protein solubility (Figure 3c,d). From untreated RPC/PS blends, the highest amount of proteins could be extracted and decreased in the order RPC/HAMS, RPC/WPS and RPC/MS. The protein solubility of untreated RPC/PS blends was equivalent to untreated RPC at pH 7 (Figure 2b), whereas blends with WPS, MS and HAMS exhibited lower protein solubility compared to RPC (Figure 3c,d).

This indicates that PS constitutes the least soluble starch among the tested varieties and accumulates as a sediment during the centrifugation step of the extraction. Accordingly, MS may exhibit a high solubility in NaCl at the given pH [59], wherefore it contributes to a larger extent to the dry matter of the extract. This in turn decreases the protein content in the extract relative to other components in the dry matter. The results of our recent studies ([14,15]), where we reported a water solubility index (%) for PS, WPS, HAMS and MS of 0.50 ± 0.00, 0.75 ± 0.35, 1.0 ± 0.14 and 1.5 ± 0.21, support these assumptions.

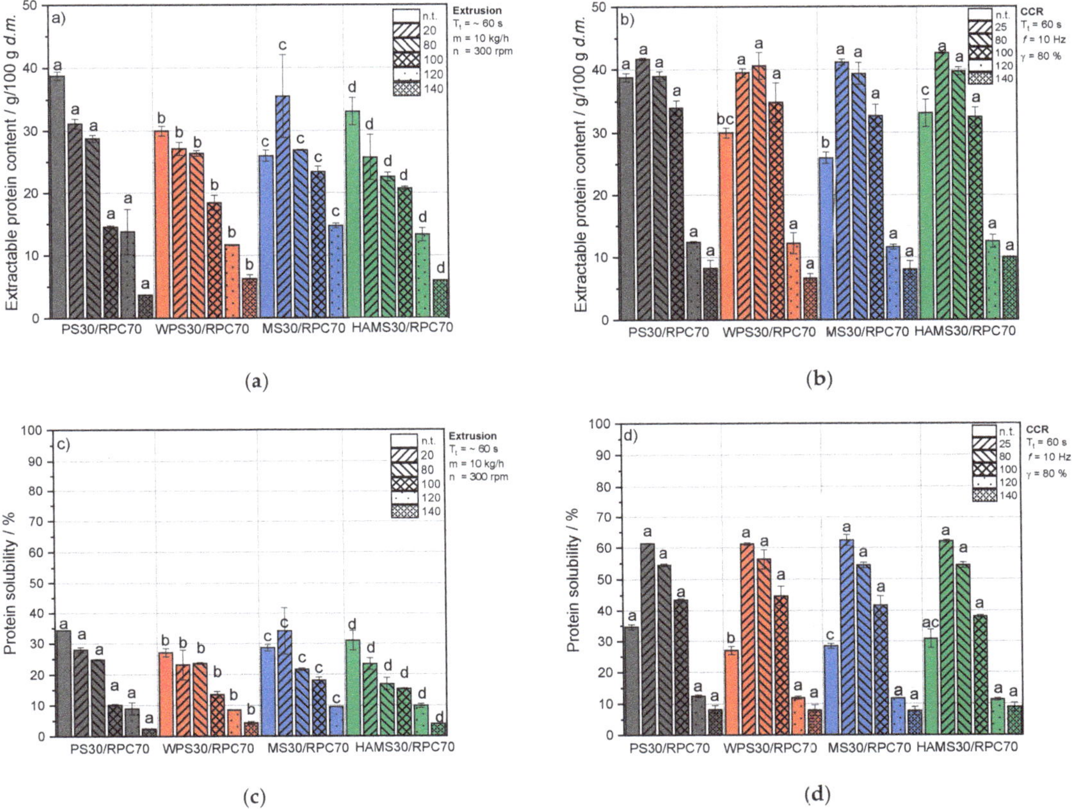

(a) (b)

(c) (d)

Figure 3. Extractable protein content (g/100 g d.m.) and protein solubility (%) at pH 7 as a function of (**a**,**c**) barrel temperature T_B (20, 80, 100, 120 or 140 °C) and (**b**,**d**) treatment temperature T_T (25, 80, 100, 120, 140) of starch/RPC blends containing 70 g/100 g w.b. RPC and 30 g/100 g w.b. potato starch (PS), waxy potato starch (WPS), maize starch (MS) or high-amylose maize starch (HAMS). Mean values with different superscript letters indicate significant differences ($p < 0.05$) between the starch types based on a two-way analysis of variance (ANOVA). Where appropriate, the mean values were compared using Tukey's honest significance test. n.t. = not treated.

Extruded Blends

In extruded blends, the starch type had a significant impact on the extractability and solubility of proteins in all blends and at all given T_B (Figure 3a,c).

At 20 °C T_B, RPC/PS and RPC/MS blends exhibited a higher protein solubility compared to RPC, whereas RPC/WPS and RPC/HAMS were equivalent to RPC. At a T_B of 80 °C, the protein solubility in the blends decreased in the order PS, WPS, MS and HAMS, with MS/RPC being equivalent to RPC as a pure component. At 100 °C T_B, the protein solubility in all potato starch blends significantly decreased compared to lower temperatures, whereas RPC and maize starch blends did not show a significant decrease in protein solubility due to this temperature increase. A T_B of 120 °C led to equivalent protein solubilities of the RPC/starch blends compared to RPC, irrespective of starch type.

Extrusion processing at T_B of 20 °C decreased the extractable protein content and solubility of PS/RPC, WPS/RPC and HAMS/RPC and a T_B increase from 20 to 80 °C, with only slightly decreased protein extractability. In MS/RPC blends, the amount of extractable and soluble protein increased when the blends were extruded at 20 °C, but decreased as a function of increasing T_B.

In the potato starch blends PS/RPC and WPS/RPC, a T_B increase from 80 to 100 °C significantly reduced the extractable protein content. This effect was greatest for PS/RPC, followed by WPS, but was only seen to a smaller extent in the maize starch blends MS/RPC and HAMS/RPC.

PS and WPS exhibit a significantly lower water-binding capacity compared to MS and HAMS, as reported in our previous studies [14,15]. This can indicate that due to severe competition on water between starches and proteins for physicochemical processes, MS and HAMS require major amounts of available process water, and can therefore limit access of water for the rapeseed proteins. Therefore, the addition of PS and WPS facilitates protein denaturation of rapeseed proteins, as indicated by a significant decrease in the protein solubility compared to MS and HAMS blends.

In extruded RPC/MS, RPC/WPS and RPC/HAMS blends, the amount of extractable protein and solubility decreased with an increase of T_B from 100 to 120 or to 140 °C. Compared to that, a T_B increase from 100 to 120 °C reduced the extractability of proteins in RPC/PS blends to a smaller extent.

Thermomechanical Treatment under Defined Conditions

Whereas the addition of starch to RPC had an effect on the protein solubility of extruded blends, almost no effect was detected when samples were treated under defined thermomechanical conditions.

However, at 100 °C, the protein solubility of blends containing PS, WPS and MS led to a slightly higher protein solubility compared to RPC and RPC/HAMS, and at 140 °C T_T the blends exhibited a slightly lower protein solubility than RPC as a pure component.

In contrast to untreated RPC/starch blends, the starch type had no significant impact on the extractability of protein in thermomechanically treated blends at all given T_T (Figure 3b,d).

With defined thermomechanical treatment at 25 °C, the amount of extractable protein increased significantly compared to untreated blends, irrespective of starch type (Figure 3b,d). A T_T increase from 25 to 80 °C slightly decreased the amount of extractable protein and solubility in all blends, and in blends with WPS a small increase of extractable protein was observed. Increasing T_T from 80 to 100 °C resulted in a decrease of extractable protein content and solubility in all blends.

When T_T increased from 100 to 120 °C, all blends exhibited significantly lower amounts of extractable protein compared to untreated blends. A further increase of T_T from 120 °C to 140 °C resulted again in a decrease of extractable protein and solubility. However, the extent was not as large as when T_T increased from 100 to 120 °C. The same effect was observed for thermomechanically treated RPC at pH 7 (Figure 2b), and can again be linked to the denaturation temperatures of cruciferin and napin [50].

Comparison of Extrusion-Processing and Defined Thermomechanical Conditions

Up to a T_T or T_B of 100 °C, blends treated under defined conditions exhibited higher amounts of extractable proteins and a higher solubility compared to extruded blends, as seen for the protein solubility of RPC (Figures 1b and 2b). The same effect was found at a T_T and T_B of 140 °C. However, at 120 °C T_T, extruded blends (Figure 3a,c) and blends that were treated under defined thermomechanical conditions (Figure 3b,d) exhibited similar amounts of solubilized protein.

In general, the shear rates that are applied onto the melt in the flow field of the extruder exhibit an inhomogeneous distribution due to the complex geometries of intermeshing twin-screw extruders [42]. However, the shear rate estimates made for this extruder type (see Table 1) were significantly higher than the shear rates applied in the closed-cavity rheometer. Therefore, it can be assumed that due to the high local share rates, high local temperatures are generated during extrusion-processing, inducing denaturation, rearrangement, hydrolysation or cross-linking. It can be assumed that the residence time of the blends at elevated temperatures were higher in the closed-cavity rheometer compared to the extruder, where not all barrel segments were heated to the same temperatures. However, the effect of high shear stress, and high temperatures at even a short residence time in the extruder barrel may have a larger impact on the denaturation, aggregation or formation of new protein linkages in RPC as does defined thermomechanical treatment at a constant high temperature, but with lower shear rates.

Moreover, the non-significant impact of starch type on the protein extraction and solubility in the RPC/starch blends treated under defined thermomechanical conditions can be taken as an indication that no protein–starch interactions are formed due to shear and heat under the given conditions. It is more likely that the effect of starch type on the protein extraction and solubility of extruded RPC/starch blends is due to the indistinct process conditions during extrusion, such as shear rate, diffuse mixing, viscosity, and as a consequence of that, the effect of heat transfer, which affects the denaturation, aggregation or formation of protein linkages.

3.4.2. Extractable Protein Content of RPC/Starch Blends at Acidic and Alkaline pH

Figure 4 illustrates the content of extractable protein in extruded RPC/starch blends as a function of starch type, T_B and pH.

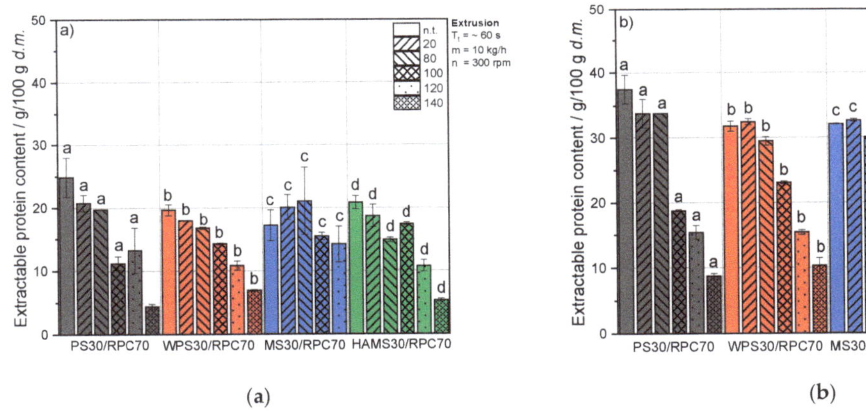

(a)

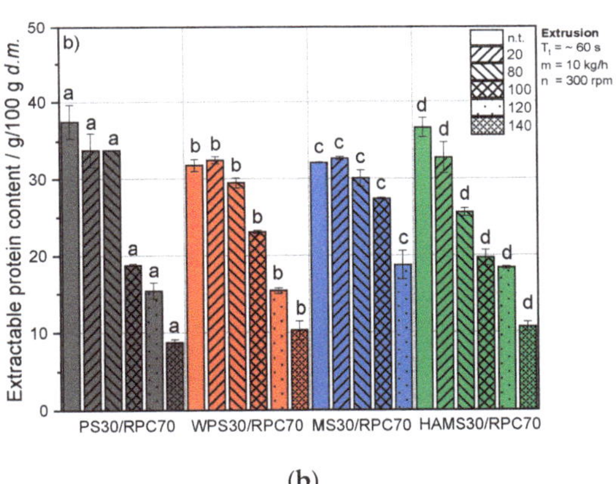

(b)

Figure 4. Extractable protein content (g/100 g d.m.) and protein solubility (%) at (**a**) pH 4, (**b**) pH 11 as a function of barrel temperature T_B (20, 80, 100, 120 or 140 °C) of starch/RPC blends containing 70 g/100 g w.b. RPC and 30 g/100 g w.b. potato starch (PS), waxy potato starch (WPS), maize starch (MS) or high-amylose maize starch (HAMS). Mean values with different superscript letters indicate significant differences ($p < 0.05$) between the starch types based on a two-way analysis of variance (ANOVA). Where appropriate, the mean values were compared using Tukey's honest significance test. n.t. = not treated.

At pH 4, T_B and the starch type had a significant impact on the protein content that was solubilized after extrusion ($p < 0.05$); however, there was no significant interaction between starch type and T_B ($p < 0.05$), and the amount of soluble protein was lower as at pH 7 and 11 (Figures 3b and 4b).

For PS, the amount of soluble protein differed only slightly between samples extruded at a T_B of 20 or 80 °C, and a large decrease of protein solubility was observed when T_B increased from 80 to 100 °C. This applied to both pH values. For MS at pH 11, the protein solubility for samples extruded at 80 and 100 °C was only slightly lower than for samples extruded at 20 °C. These two observations can be related to the low T_{gel} of PS and the comparatively high T_{gel} of MS (discussed in Section 4).

3.5. Impact of Extrusion on Starch Gelatinization and Hydrolysis

Table 4 reports the impact of barrel temperature and starch type on the gelatinization temperature, degree and enthalpy of extruded RPC/starch blends. T_B had a significant impact on T_{gel} ($p < 0.05$) and ΔH_{gel} ($p < 0.01$) of all starches, the starch types differed significantly in ΔH_{gel} ($p < 0.01$), but did not differ significantly in T_{gel} ($p > 0.05$). After extrusion, the thermal properties (ΔH_{gel}) of the blends were significantly lower as in the unextruded samples.

Table 4. Degree of gelatinization D_{gel}, gelatinization temperature T_{gel} and gelatinization enthalpy ΔH_{gel} of 70 g/100 g rapeseed press cake (RPC) combined with 30 g/100 g potato starch (PS), waxy potato starch (WPS), maize starch (MS) and high-amylose maize starch (HAMS).

Barrel Temperature T_B (°C)	PS30/RPC70	WPS30/RPC70	MS30/RPC70	HAMS30/RPC70
		Degree of gelatinization D_{gel} (%)		
20	58.14 [a]	49.59 [b]	81.71 [c]	No peak
80	78.11 [a]	93.31 [b]	81.68 [c]	No peak
100	94.97 [a]	No peak	56.26 [b]	No peak
120	No peak	No peak	91.39	No peak
140	No peak	No peak	No peak	No peak
		Gelatinization temperature T_{gel} (°C)		
Not extruded	68.82 ± 0.49 [a]	54.29 ± 0.51 [b]	67.65 ± 0.48 [c]	No peak
20	69.48 ± 1.08 [a]	59.15 ± 0.49 [b]	70.92 ± 1.24 [c]	No peak
80	67.85 ± 0.14 [a]	56.87 ± 0.12 [b]	70.84 ± 1.01 [c]	No peak
100	67.14 ± 0.11 [a]	No peak	70.74 ± 0.88 [b]	No peak
120	No peak	No peak	73.35 ± 0.64	No peak
140	No peak	No peak	No peak	No peak
		Gelatinization enthalpy ΔH_{gel} (J/g)		
Not extruded	1.74 ± 0.07 [a]	8.12 ± 0.23 [b]	1.16 ± 0.19 [c]	No peak
20	0.73 ± 0.17 [a]	4.09 ± 0.12 [b]	0.21 ± 0.01 [c]	No peak
80	0.38 ± 0.00 [a]	0.54 ± 0.01 [b]	0.21 ± 0.01 [c]	No peak
100	0.08 ± 0.01 [a]	No peak	0.50 ± 0.01 [b]	No peak
120	No peak	No peak	0.09 ± 0.01	No peak
140	No peak	No peak	No peak	No peak

Mean values with different superscript letters within one row indicate significant differences ($p < 0.05$) based on a one-way analysis of variance (ANOVA). Where appropriate, the mean values were compared using Tukey's honest significance test.

WPS blends exhibited the lowest T_{gel} and the highest ΔH_{gel} compared to PS and MS blends. D_{gel} increased with increasing T_B and a D_{gel} of > 90% was found for WPS blends at a T_B of 80 °C, for PS blends at 100 °C and for MS blends at 120 °C. PS30/RPC70 had a higher T_{gel} than WPS30/RPC70, but though lower than MS30/RPC70. This explains that for PS and MS blends, the stage of full gelatinization, indicated by the absence of a peak in the thermogram, was completed at 100 °C and 120 °C, respectively. The ΔH_{gel} of PS, WPS and MS blends had already decreased by 58, 50 and 81% when the blends were extruded at 20 °C compared to unextruded blends. For HAMS, no thermal peak was observed at any T_B.

Starch gelatinization indicates the disorganization of the semi-crystalline structure into an amorphous state, and since HAMS exhibits >67% amorphous amylose and the blends only contained 30 g/100 g (w.b.) starch, the transformation of the remaining crystalline amylopectin into the amorphous state in the blends might not be detectable with the DSC. In a number of previous studies, there was also no distinct endothermic peak found for the gelatinization of high-amylose starches. Russel et al. [60] detected one broad endotherm between 66 and 104 °C during the heating of amylomaize starch containing 70% amylose conditioned to 57% water. Similar observations were made by Eberstein et al. [61]. The authors considered the DSC to be not sufficiently sensitive for the detection of gelatinization.

The high ΔH_{gel} of WPS30/RPC70, and the low T_{gel} compared to other starch types is also in accordance with the literature and can be attributed to a high water uptake and a high degree of transformation from crystalline to amorphous structures of amylopectin rich starches [30,62]. In contrast, high-amylose starches are known to exhibit a low water absorption and solubility, and an overall high resistance to gelatinization and hydrolysis [63,64]. This is in alignment with our study, where no thermal transition peak was present in HAMS30/RPC70 [65,66].

An increase in D_{gel} with an increasing extrusion temperature was reported several times in the literature for sweet potato starch [67] and maize starch [68] as single components, as well as for multicomponent biopolymers, such as bran-enriched wheat flour [69].

In the RPC/starch blends, severe competition for water likely takes place with both components requiring water for physicochemical transformations, especially since the starch and protein content in the blends was set to be relatively equal (27 and 30 g/100 g respectively). Furthermore, the fibers present in the RPC have a high capacity to hydrate, wherefore they restrict the availability of the plasticizer, increase the melt viscosity, and reduce the availability of water required for gelatinization [69]. Although there was an excess of RPC in the blends and the process water was limited to 29 g/100 g d.m., full starch gelatinization for PS, WPS and MS was observed at max. 120 °C.

As illustrated in Figure 5, the starch type and T_B had a significant impact on S_H; furthermore, a significant interaction of starch type and T_B was observed ($p < 0.01$). In the unextruded blends, almost the whole amount of PS, WPS and MS that was present in the blends (30 g/100 g), was hydrolysed, wherefore only 27 g/100 g HAMS could be hydrolyzed. MS in extruded blends at a T_B of 140 °C could almost fully be hydrolyzed, whereas for PS, WPS and HAMS, S_H decreased with increasing T_B. A T_B of 140 °C only marginally decreases S_H for MS, WPS and HAMS blends, and slightly increases S_H for the PS blend, compared to a T_B of 100 °C.

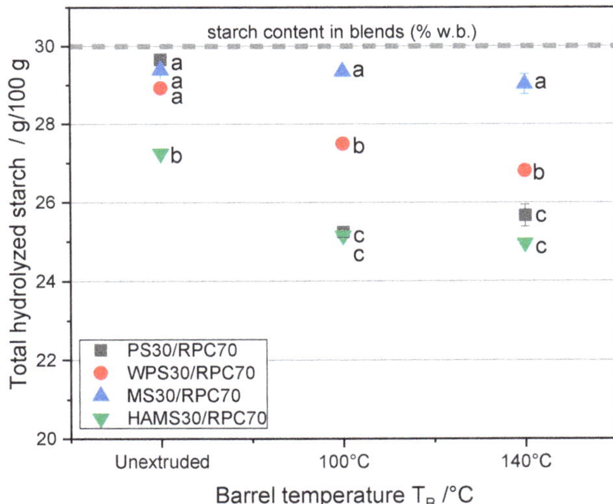

Figure 5. Content of total hydrolyzed starch (S_H, g/100 g) of 70 g/100 g rapeseed press cake (RPC) mixed with 30 g/100 g potato starch (PS), waxy potato starch (WPS), maize starch (MS) and high-amylose maize starch (HAMS). Mean values with different superscript letters indicate significant differences ($p < 0.05$) between the starch types based on a two-way analysis of variance (ANOVA). Where appropriate, the mean values were compared using Tukey's honest significance test.

A decrease of S_H can indicate the formation of non-covalent or covalent bonds between starch and other components in the blends, such as rapeseed proteins, lipids or fibers, during extrusion [26]. Extrusion-induced starch–protein interactions have particularly been investigated in previous studies, though markedly with a focus on whey proteins [31,70,71]. Those effects can be detected by a parallel investigation of protein solubility and starch hydrolysis.

Allen et al. [31] discussed three possible reasons for reduced protein solubility when proteins are extruded in starch blends. The small protein molecules might be physically entrapped in the amylopectin matrix that is partially broken down during gelatinization. Since in our study, the reduction of protein solubility was not larger in WPS blends than in PS, MS or HAMS blends, we assume that this effect did not occur. Additionally, the authors discussed the transformation of a crystalline to a more amorphous structure, and the leakage of amylose out of the granula might provide the opportunity for proteins to align with amylose molecules in the shear zone of the extruder and stabilize due to covalent bonds after exiting the die [29,31,72]. However, the solubility of proteins extruded with HAMS (67% amylose) was not lower compared to the starches containing a higher content of amylopectin, therefore with our results, this effect could not be observed. A third theory is the formation of covalent protein–starch linkages during thermomechanical treatment, that would decrease protein solubility and the amount of hydrolyzed starch synergistically [31,72]. Since the amount of hydrolyzed starch did only slightly decrease in MS blends and to a slightly higher extent in WPS and HAMS blends, when T_B increased from 100 to 140 °C, but the protein solubility decreased significantly in these samples, we assume, that rather than protein–starch bonds, protein–protein bonds between unfolded rapeseed proteins are formed during extrusion that cannot be solubilized in NaCl.

The lower degree of hydrolysis of HAMS-based blends compared to MS, PS and WPS blends, regardless of whether the samples were extruded or not, may be attributed to the low susceptibility to degradation of HAMS due to its high amylose content. Some studies report that high-amylose maize starch is less susceptible to various physicochemical treatments (e.g., hydrothermal treatments) than normal or waxy starches, due to its lower crystallinity, small particle size and high surface area [66,73–76].

In our study, the extrusion-induced reduction of hydrolyzed starch content in PS30/RPC70 may indicate that the starch has interacted with rapeseed components during extrusion, making the starch less accessible for the enzymes during analysis. This observation could be corroborated by the relatively low T_{gel} and the comparatively large decrease of protein solubility at T_B 100 °C of PS30/RPC70, indicating that polymerization occurred from interactions between exposed binding sites from unfolded rapeseed proteins and gelatinized starch. However, since process conditions during extrusion processing were significantly influenced by the starch type, and this effect was not found in blends treated under defined thermomechanical conditions, we assume that rheological effects due to the starch type during the analysis of starch and protein solubility dominated over actual protein–starch bonds. To resolve this, future studies should systematically evaluate the impact of an isolated rapeseed protein addition to starch blends under defined thermomechanical conditions.

3.6. Impact of Extrusion on Total, Soluble, and Insoluble Dietary Fiber

Figure 6 shows the ratio of soluble to total dietary fiber in RPC and RPC/starch blends as a function of T_B.

RPC exhibited the least amount of SDF relative to TDF in an unextruded and extruded condition. With the addition of starch, SDF/TDF increased in the order of RPC/HAMS, RPC/MS, RPC/WPS and RPC/PS. With an increase of T_B to 100 °C, the SDF/TDF of RPC increased, but an increase to 120 °C did not affect the SDF/TDF of RPC.

With a T_B of 100 and 140 °C, the amount of SDF increased regardless of the starch type in the blends. There was only a slight increase of SDF in the samples when T_B increased from 100 to 140 °C; however, SDF/TDF of RPC/HAMS increased to a larger extent.

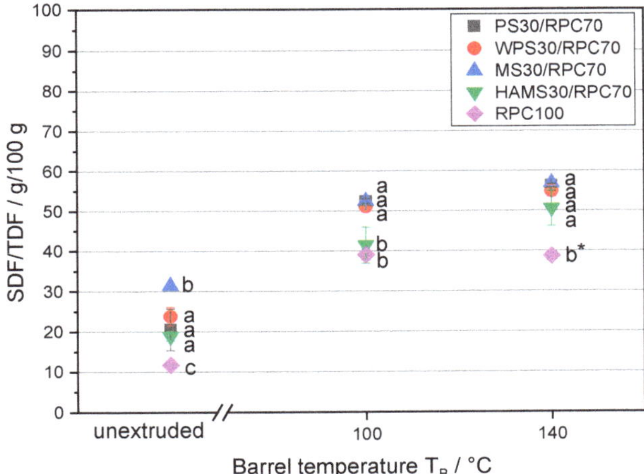

Figure 6. Ratio of soluble (SDF) to total (TDF) dietary fiber (SDF/TDF) of rapeseed press cake (RPC100) and RPC/starch blends containing potato starch (PS), waxy potato starch (WPS), maize starch (MS) or high-amylose maize starch (HAMS) as a function of barrel temperature T_B (unextruded, 100 °C, 140 °C). Mean values with different superscript letters indicate significant differences ($p < 0.05$) between the blends based on a two-way analysis of variance (ANOVA). Where appropriate, the mean values were compared using Tukey's honest significance test. * Extruded at 120 °C.

The increase of SDF/TDF with the addition of starch to RPC can be related to the methodology of dietary fiber analysis in our study, where the first step of sample treatment is an enzymatic digestion of amylose using amylase. The presence of antinutritional components in RPC, such as phytic acid, can inhibit the starch digestion by amylase, as shown in previous research [50,77]. Furthermore, the presence of resistant starch can increase the measured SDF values, since they are not available for enzymatic hydrolysis by amylase. Blends containing HAMS consequently exhibit the least amount of soluble, and the highest amount of insoluble fractions. This effect was also seen for starch hydrolysis (Figure 5). HAMS can be considered as resistant starch, less affected by enzymatic hydrolysis than high amylopectin starches like WPS.

A number of studies based on rice bran or wheat bran described an increase of SDF/TDF due to extrusion [32,34]. An extrusion caused by a shift from insoluble to more soluble dietary fiber fractions has recently been observed by Naumann et al. [33], when lupin kernel fiber was processed in a laboratory twin-screw extruder, accompanied by a large increase of water-binding capacity. The authors associated the redistribution of IDF to SDF to mechanical rather than to thermal effects, in accordance to the studies of Ralet et al. [78].

An increase of SDF/TDF was accompanied by a high SME in the study of Naumann et al. [33]. However, in our study WPS exhibited the highest SME at a T_B of 100 °C and 140 °C, but the increase of SDF/TDF was not larger than in MS or HAMS, which generated lower SMEs during extrusion. The slight increase of SDF when T_B increased from 100 to 140 °C can be explained by the impact of thermal treatment on additional breaks of glycosidic bonds resulting in smaller and more soluble fractions of polysaccharides, as described in previous studies [79].

4. Conclusions

In this study, rapeseed press cake (RPC), both pure and in combination with four starch types, and varying in botanical origin and amylose content, was exposed to thermomechanical treatment at five temperatures. The impact of extrusion-processing on the solubility of rapeseed protein and fiber, hydrolyzed starch content and gelatinization properties was investigated using extraction, enzymatic and enzymatic-gravimetric analysis. Starch and protein content in the blends was equal in order to exclude an overage effect.

Indistinct process conditions (product temperatures and specific mechanical energy input) were monitored during extrusion as a function of starch type. To compensate this effect, a closed-cavity rheometer was used to expose RPC and the blends to defined thermomechanical treatment at a constant shear rate and extrusion-like temperatures. The protein solubility of the thermomechanically treated blends was analyzed accordingly.

Extrusion-processing of RPC significantly reduced the protein solubility with increasing barrel temperature, so did thermomechanical treatment at defined conditions with increasing treatment temperatures. At temperatures $\leq$100 °C, extruded RPC exhibited a lower protein solubility compared to thermomechanically treated RPC, whereas both treatments led to an equivalent protein solubility at 120 °C. Effects observed at temperatures $\leq$100 °C can be attributed to the higher shear forces and higher local temperatures in the extruder barrel promoting protein aggregation reactions. At $\geq$100 °C, it can be assumed that thermal energy input dominates over the influence of shear stress, wherefore extrusion and defined thermomechanical treatment resulted in equivalent protein solubilities.

The protein solubility of extruded RPC/starch blends was significantly influenced by the starch type and decreased with increased barrel temperature. No effect of starch type was detected when samples were treated under defined thermomechanical conditions. This was considered as an indication that no protein–starch interactions were formed due to shear and heat under the given conditions. It can be assumed that the effect of starch type on the protein solubility in extruded blends was predominantly due to differences in the rheological properties of the starches, leading to indistinct process conditions during extrusion, such as shear rates or local temperatures.

At temperatures of 100 and 140 °C, blends treated with defined thermomechanical conditions exhibited higher amounts of extractable proteins and a higher protein solubility compared to extruded blends, which was attributed to higher local shear rates in the extruder barrel compared to the shear rates applied in the rheometer.

Increasing barrel temperatures decreased the degree of hydrolyzed starch, and increased the degree of gelatinization, as well as the amount of soluble dietary fibers of RPC; however, the latter was not affected by the starch type.

Our findings emphasize that the closed-cavity rheometer is a suitable tool to analyze physicochemical transformations of biopolymers at defined thermomechanical conditions in order to overcome the hurdle of indistinct process conditions during extrusion, as they are significantly affected by the raw material.

Author Contributions: A.M.: data curation, writing—original draft preparation, formal analysis, visualization. S.N.: data curation, writing—review and editing. R.O.: writing—review and editing. H.P.K.: writing—review and editing. M.A.E.: supervision, conceptualization, methodology, writing—review and editing. All authors have read and agreed to the published version of the manuscript.

Funding: This research received no external funding.

Institutional Review Board Statement: Not applicable.

Informed Consent Statement: Not applicable.

Data Availability Statement: No data reported.

Acknowledgments: The authors would like to thank Sigrid Gruppe, Elfriede Bischof and Evi Muller for the chemical analyses and Michael Schott for the particle size analysis. The authors are grateful to Natalia Garza and Naomi Albiez for her help in the extrusion processing and analysis of the extrudates. Additionally, the authors thank Isabel Muranyi for her support in data analysis.

Conflicts of Interest: The authors declare no conflict of interest.

Abbreviations

D_{gel}	Degree of gelatinization (%)
d.m.	Dry matter (g/100g)
HAMS	High-amylose maize starch
ΔH_{gel}	Gelatinization enthalpy (J/g)
IDF	Insoluble dietary fiber (g/100 g)
m	Mass flow rate (kg/h)
MS	Maize starch
n	Screw speed (rpm)
n.t.	Not treated
PS	Potato starch
RPC	Rapeseed press cake
SDF	Soluble dietary fiber (g/100 g)
S_H	Total hydrolyzed starch (g/100 g)
SME	Specific mechanical energy (Wh/kg)
T	Temperature (°C)
T_B	Barrel temperature (°C)
TDF	Total soluble dietary fiber (g/100 g)
T_{gel}	Gelatinization temperature (°C)
T_T	Treatment temperature (°C)
T_t	Treatment time (s)
w.b.	Wet basis (g/100 g)
WPS	Waxy potato starch

References

1. Hoglund, E.; Eliasson, L.; Oliveira, G.; Almli, V.L.; Sozer, N.; Alminger, M. Effect of drying and extrusion processing on physical and nutritional characteristics of bilberry press cake extrudates. *LWT Food Sci. Technol.* **2018**, *92*, 422–428. [CrossRef]
2. Yagci, S.; Gogus, F. Response surface methodology for evaluation of physical and functional properties of extruded snack foods developed from food-by-products. *J. Food Eng.* **2008**, *86*, 122–132. [CrossRef]
3. Da Silva, A.P.L.; Berrios, J.D.J.; Pan, J.; Ascheri, J.L.R. Passion fruit shell flour and rice blends processed into fiber-rich expanded extrudates. *CyTA J. Food* **2018**, *16*, 901–908. [CrossRef]
4. Karkle, E.L.; Alavi, S.; Dogan, H. Cellular architecture and its relationship with mechanical properties in expanded extrudates containing apple pomace. *Food Res. Int.* **2012**, *46*, 10–21. [CrossRef]
5. Wang, S.Y.; Kowalski, R.J.; Kang, Y.F.; Kiszonas, A.M.; Zhu, M.J.; Ganjyal, G.M. Impacts of the Particle Sizes and Levels of Inclusions of Cherry Pomace on the Physical and Structural Properties of Direct Expanded Corn Starch. *Food Bioprocess. Tech.* **2017**, *10*, 394–406. [CrossRef]
6. Makila, L.; Laaksonen, O.; Diaz, J.M.R.; Vahvaselka, M.; Myllymaki, O.; Lehtomaki, I.; Laakso, S.; Jahreis, G.; Jouppila, K.; Larmo, P.; et al. Exploiting blackcurrant juice press residue in extruded snacks. *LWT Food Sci. Technol.* **2014**, *57*, 618–627. [CrossRef]
7. Onwulata, C.I.; Konstance, R.P. Extruded corn meal and whey protein concentrate: Effect of particle size. *J. Food Process. Pres.* **2006**, *30*, 475–487. [CrossRef]
8. Onwulata, C.I.; Smith, P.W.; Konstance, R.P.; Holsinger, V.H. Incorporation of whey products in extruded corn, potato or rice snacks. *Food Res. Int.* **2001**, *34*, 679–687. [CrossRef]
9. Robin, F.; Dubois, C.; Pineau, N.; Schuchmann, H.P.; Palzer, S. Expansion mechanism of extruded foams supplemented with wheat bran. *J. Food Eng.* **2011**, *107*, 80–89. [CrossRef]
10. Nikinmaa, M.; Nordlund, E.; Poutanen, K.; Sozer, N. From Underutilized Side-Streams to Hybrid Food Ingredients for Health. *Cereal Food World* **2018**, *63*, 137–142. [CrossRef]
11. Martin, A.; Osen, R.; Greiling, A.; Karbstein, H.P.; Emin, A. Effect of of rapeseed press cake and peel on the extruder response and physical pellet quality in extruded fish feed. *Aquaculture* **2019**, *512*. [CrossRef]
12. Tyapkova, O.; Osen, R.; Wagenstaller, M.; Baier, B.; Specht, F.; Zacherl, C. Replacing fishmeal with oilseed cakes in fish feed—A study on the influence of processing parameters on the extrusion behavior and quality properties of the feed pellets. *J. Food Eng.* **2016**, *191*, 28–36. [CrossRef]
13. Smulikowska, S.; Czerwiński, J.; Mieczkowska, A. Nutritional value of rapeseed expeller cake for broilers: Effect of dry extrusion. *J. Anim. Feed Sci.* **2006**, *15*, 445–453. [CrossRef]
14. Martin, A.; Osen, R.; Karbstein, H.P.; Emin, M.A. Impact of Rapeseed Press Cake on the Rheological Properties and Expansion Dynamics of Extruded Maize Starch. *Foods* **2021**, *10*, 616. [CrossRef] [PubMed]
15. Martin, A.; Osen, R.; Karbstein, H.P.; Emin, M.A. Linking Expansion Behaviour of Extruded Potato Starch/Rapeseed Press Cake Blends to Rheological and Technofunctional Properties. *Polymers* **2021**, *13*, 215. [CrossRef] [PubMed]

16. Khattab, R.Y.; Arntfield, S.D. Functional properties of raw and processed canola meal. *LWT Food Sci. Technol.* **2009**, *42*, 1119–1124. [CrossRef]
17. Mahajan, A.; Dua, S. Nonchemical approach for reducing antinutritional factors in rapeseed (Brassica campestris var. Toria) and characterization of enzyme phytase. *J. Agric. Food Chem.* **1997**, *45*, 2504–2508. [CrossRef]
18. Mahajan, A.; Dua, S.; Bhardwaj, S. Imbibition induced changes in antinutritional constituents and functional properties of rapeseed (Brassica campestris var. toria) meal. *FASEB J.* **1997**, *11*, A1105-A1105.
19. Mawson, R.; Heaney, R.K.; Zdunczyk, Z.; Kozlowska, H. Rapeseed Meal-Glucosinolates and Their Antinutritional Effects.6. Taint in End-Products. *Nahrung* **1995**, *39*, 21–31. [CrossRef]
20. Lomascolo, A.; Uzan-Boukhris, E.; Sigoillot, J.C.; Fine, F. Rapeseed and sunflower meal: A review on biotechnology status and challenges. *Appl. Microbiol. Biot.* **2012**, *95*, 1105–1114. [CrossRef]
21. DellaValle, G.; Colonna, P.; Patria, A.; Vergnes, B. Influence of amylose content on the viscous behavior of low hydrated molten starches. *J. Rheol.* **1996**, *40*, 347–362. [CrossRef]
22. Fetzer, A.; Herfellner, T.; Stabler, A.; Menner, M.; Eisner, P. Influence of process conditions during aqueous protein extraction upon yield from pre-pressed and cold-pressed rapeseed press cake. *Ind. Crop. Prod.* **2018**, *112*, 236–246. [CrossRef]
23. Leming, R.; Lember, A. Chemical composition of expeller-extracted and cold-pressed rapeseed cake. *Agraarteadus* **2005**, *16*, 103–109.
24. Ancuța, P.; Sonia, A. Oil Press-Cakes and Meals Valorization through Circular Economy Approaches: A Review. *Appl. Sci.* **2020**, *10*, 7432. [CrossRef]
25. Pastor-Cavada, E.; Drago, S.R.; González, R.J.; Juan, R.; Pastor, J.E.; Alaiz, M.; Vioque, J. Effects of the addition of wild legumes (Lathyrus annuus and Lathyrus clymenum) on the physical and nutritional properties of extruded products based on whole corn and brown rice. *Food Chem.* **2011**, *128*, 961–967. [CrossRef]
26. Day, L.; Swanson, B.G. Functionality of Protein-Fortified Extrudates. *Compr. Rev. Food Sci. Food Saf.* **2013**, *12*, 546–564. [CrossRef] [PubMed]
27. Quevedo, M.; Jandt, U.; Kulozik, U.; Karbstein, H.P.; Emin, M.A. Investigation on the influence of high protein concentrations on the thermal reaction behaviour of beta-lactoglobulin by experimental and numerical analyses. *Int. Dairy J.* **2019**, *97*, 99–110. [CrossRef]
28. Zhang, B.; Liu, G.; Ying, D.; Sanguansri, L.; Augustin, M.A. Effect of extrusion conditions on the physico-chemical properties and in vitro protein digestibility of canola meal. *Food Res. Int.* **2017**, *100*, 658–664. [CrossRef]
29. Matthey, F.P.; Hanna, M.A. Physical and Functional Properties of Twin-screw Extruded Whey Protein Concentrate–Corn Starch Blends. *Lebensm Wiss Technol.* **1997**, *30*, 359–366. [CrossRef]
30. Zhang, W.; Li, S.; Zhang, B.; Drago, S.R.; Zhang, J. Relationships between the gelatinization of starches and the textural properties of extruded texturized soybean protein-starch systems. *J. Food Eng.* **2016**, *174*, 29–36. [CrossRef]
31. Allen, K.E.; Carpenter, C.E.; Walsh, M.K. Influence of protein level and starch type on an extrusion-expanded whey product. *Int. J. Food Sci. Technol.* **2007**, *42*, 953–960. [CrossRef]
32. Andersson, A.A.M.; Andersson, R.; Jonsäll, A.; Andersson, J.; Fredriksson, H. Effect of Different Extrusion Parameters on Dietary Fiber in Wheat Bran and Rye Bran. *J. Food Sci. J. Food Sci.* **2017**, *82*, 1344–1350. [CrossRef] [PubMed]
33. Naumann, S.; Schweiggert-Weisz, U.; Martin, A.; Schuster, M.; Eisner, P. Effects of extrusion processing on the physiochemical and functional properties of lupin kernel fibre. *Food Hydrocoll.* **2021**, *111*, 106222. [CrossRef]
34. Vasanthan, T.; Gaosong, J.; Yeung, J.; Li, J. Dietary fiber profile of barley flour as affected by extrusion cooking. *Food Chem.* **2002**, *77*, 35–40. [CrossRef]
35. German Food Act. Methods L. 16.01-2, L. 17.00-1, L. 17.00-3. In *BVL Bundesamt für Verbraucherschutz und Lebensmittelsicherheit*; Beuth Verlag GmbH: Berlin, Germany, 2005.
36. AOAC International. *Official Method 945.46—Ash Determination*; The Scientific Association Dedicated to Analytical Excellence: Washington, DC, USA, 2016.
37. AOAC International. *Official Method 962.09—Fiber (Crude) in Animal Feed and Pet. Food*; The Scientific Association Dedicated to Analytical Excellence: Washington, DC, USA, 1982.
38. Beutler, H.O. Enzymatic Determination of Starch in Foods by Hexokinase Method. *Starch Stärke* **1978**, *30*, 309–312. [CrossRef]
39. AACC. *Hydration Capacity of Pregelatinized Cereal Products. Approved Methods of Analysis*, 11th ed.; AACCI Method 56-20.01; American Association of Cereal Chemists International: St. Paul, MN, USA, 1999.
40. Osen, R.; Toelstede, S.; Wild, F.; Eisner, P.; Schweiggert-Weisz, U. High moisture extrusion cooking of pea protein isolates: Raw material characteristics, extruder responses, and texture properties. *J. Food Eng.* **2014**, *127*, 67–74. [CrossRef]
41. Koch, L.; Emin, M.A.; Schuchmann, H.P. Reaction behaviour of highly concentrated whey protein isolate under defined heat treatments. *Int. Dairy J.* **2017**, *71*, 114–121. [CrossRef]
42. Emin, M.A.; Schuchmann, H.P. Analysis of the dispersive mixing efficiency in a twin-screw extrusion processing of starch based matrix. *J. Food Eng.* **2013**, *115*, 132–143. [CrossRef]
43. Liu, K.; Hsieh, F.-H. Protein-protein interactions during high-moisture extrusion for fibrous meat analogues and comparison of protein solubility methods using different solvent systems. *J. Agric. Food Chem.* **2008**, *56*, 2681–2687. [CrossRef]
44. Ebeling, M.E. The Dumas Method for Nitrogen in Feeds. *J. AOAC Int.* **1968**, *51*, 766–770. [CrossRef]
45. Maki, K.C.; Pelkman, C.L.; Finocchiaro, E.T.; Kelley, K.M.; Lawless, A.L.; Schild, A.L.; Rains, T.M. Resistant starch from high-amylose maize increases insulin sensitivity in overweight and obese men. *J. Nutr.* **2012**, *142*, 717–723. [CrossRef] [PubMed]

46. Li, M.; Hasjim, J.; Xie, F.; Halley, P.J.; Gilbert, R.G. Shear degradation of molecular, crystalline, and granular structures of starch during extrusion. *Starch Stärke* **2014**, *66*, 595–605. [CrossRef]
47. Tzeng, Y.-M.; Diosady, L.L.; Rubin, L.J. Preparation of Rapeseed Protein Isolates Using Ultrafiltration, Precipitation and Diafiltration. *Can. Inst. Food Sci. Technol. J.* **1988**, *21*, 419–424. [CrossRef]
48. Arntfield, S.D.; Murray, E.D. The Influence of Processing Parameters on Food Protein Functionality I. Differential Scanning Calorimetry as an Indicator of Protein Denaturation. *Can. Inst. Food Sci. Technol. J.* **1981**, *14*, 289–294. [CrossRef]
49. Perera, S.P.; McIntosh, T.C.; Wanasundara, J.P.D. Structural Properties of Cruciferin and Napin of Brassica napus (Canola) Show Distinct Responses to Changes in pH and Temperature. *Plants* **2016**, *5*, 36. [CrossRef]
50. Tan, S.H.; Mailer, R.J.; Blanchard, C.L.; Agboola, S.O. Canola Proteins for Human Consumption: Extraction, Profile, and Functional Properties. *J. Food Sci. J. Food Sci.* **2011**, *76*, R16–R28. [CrossRef]
51. Chen, B.Y.; Yu, C.; Liu, J.F.; Yang, Y.L.; Shen, X.C.; Liu, S.W.; Tang, X.Z. Physical properties and chemical forces of extruded corn starch fortified with soy protein isolate. *Int. J. Food Sci. Tech.* **2017**, *52*, 2604–2613. [CrossRef]
52. Salazar-Villanea, S.; Bruininx, E.M.A.M.; Gruppen, H.; Hendriks, W.H.; Carré, P.; Quinsac, A.; van der Poel, A.F.B. Physical and chemical changes of rapeseed meal proteins during toasting and their effects on in vitro digestibility. *J. Anim. Sci. Biotechnol.* **2016**, *7*, 62. [CrossRef]
53. Mosenthin, R.; Messerschmidt, U.; Sauer, N.; Carré, P.; Quinsac, A.; Schöne, F. Effect of the desolventizing/toasting process on chemical composition and protein quality of rapeseed meal. *J. Anim. Sci. Biotechnol.* **2016**, *7*, 36. [CrossRef]
54. Rommi, K.; Ercili-Cura, D.; Hakala, T.K.; Nordlund, E.; Poutanen, K.; Lantto, R. Impact of total solid content and extraction pH on enzyme-aided recovery of protein from defatted rapeseed (Brassica rapa L.) press cake and physicochemical properties of the protein fractions. *J. Agric. Food Chem.* **2015**, *63*, 2997–3003. [CrossRef]
55. Miller, N.; Pretorius, H.E.; Du Toit, L.J. Phytic acid in sunflower seeds, pressed cake and protein concentrate. *Food Chem.* **1986**, *21*, 205–209. [CrossRef]
56. Hídvégi, M.; Lásztity, R. Phytic acid content of cereals and legumes and interaction with proteins. *Period. Polytech. Chem. Eng.* **2002**, *46*, 59–64.
57. Emin, M.A.; Quevedo, M.; Wilhelm, M.; Karbstein, H.P. Analysis of the reaction behavior of highly concentrated plant proteins in extrusion-like conditions. *Innov. Food Sci. Emerg.* **2017**, *44*, 15–20. [CrossRef]
58. Wanasundara, J.P.D.; Tan, S.; Alashi, A.M.; Pudel, F.; Blanchard, C. *Proteins from Canola/Rapeseed: Current Status*; Sustainable Protein Sources; Elsevier: Amsterdam, The Netherlands, 2017; pp. 285–304.
59. Hedayati, S.; Shahidi, F.; Koocheki, A.; Farahnaky, A.; Majzoobi, M. Physical properties of pregelatinized and granular cold water swelling maize starches at different pH values. *Int. J. Biol. Macromol.* **2016**, *91*, 730–735. [CrossRef]
60. Russell, P.L. Gelatinisation of starches of different amylose/amylopectin content. A study by differential scanning calorimetry. *J. Cereal. Sci.* **1987**, *6*, 133–145. [CrossRef]
61. Eberstein, K.; Höpcke, R.; Kleve; Konieczny-Janda, G.; Stute, R. DSC-Untersuchungen an Stärke Teil I. Möglichkeiten thermoanalytischer Methoden zur Stärkecharakterisierung. *Starch Stärke* **1980**, *32*, 397–404. [CrossRef]
62. Kibar, E.A.A.; Gönenç, İ.; Us, F. Gelatinization of Waxy, Normal and High Amylose Corn Starches. *J. Food* **2010**, *35*, 237–244.
63. Ye, J.P.; Hu, X.T.; Luo, S.J.; Liu, W.; Chen, J.; Zeng, Z.R.; Liu, C.M. Properties of Starch after Extrusion: A Review. *Starch Stärke* **2018**, *70*. [CrossRef]
64. Lai, L.S.; Kokini, J.L. The effect of extrusion operating conditions on the on-line apparent viscosity of 98% Amylopectin (Amioca) and 70% Amylose (Hylon 7) corn starches during extrusion. *J. Rheol.* **1990**, *34*, 1245–1266. [CrossRef]
65. Lin, L.; Guo, D.; Zhao, L.; Zhang, X.; Wang, J.; Zhang, F.; Wei, C. Comparative structure of starches from high-amylose maize inbred lines and their hybrids. *Food Hydrocolloid* **2016**, *52*, 19–28. [CrossRef]
66. Tan, X.; Zhang, B.; Chen, L.; Li, X.; Li, L.; Xie, F. Effect of planetary ball-milling on multi-scale structures and pasting properties of waxy and high-amylose cornstarches. *Innov. Food Sci. Emerg.* **2015**, *30*, 198–207. [CrossRef]
67. Waramboi, J.G.; Gidley, M.J.; Sopade, P.A. Influence of extrusion on expansion, functional and digestibility properties of whole sweetpotato flour. *Lebensm Wiss Technol.* **2014**, *59*, 1136–1145. [CrossRef]
68. Bhatnagar, S.; Hanna, M.A. Extrusion Processing Conditions for Amylose Lipid Complexing. *Cereal Chem.* **1994**, *71*, 587–593.
69. Altan, A.; McCarthy, K.L.; Maskan, M. Effect of Extrusion Cooking on Functional Properties and in vitro Starch Digestibility of Barley-Based Extrudates from Fruit and Vegetable By-Products. *J. Food Sci. J. Food Sci.* **2009**, *74*, E77–E86. [CrossRef] [PubMed]
70. Kumar, L.; Brennan, M.A.; Mason, S.L.; Zheng, H.; Brennan, C.S. Rheological, pasting and microstructural studies of dairy protein-starch interactions and their application in extrusion-based products: A review. *Starch Stärke* **2017**, *69*, 1600273. [CrossRef]
71. Onwulata, C.I.; Tunick, M.H.; Thomas-Gahring, A.E. Pasting and Extrusion Properties of Mixed Carbohydrate and Whey Protein Isolate Matrices. *J. Food Process. Pres.* **2014**, *38*, 1577–1591. [CrossRef]
72. Kim, C.H.; Maga, J.A. Properties of Extruded Whey Protein Concentrate and Cereal Flour Blends. *Lebensm. Wiss. Technol.* **1987**, *20*, 311–318UR. Available online: https://pascal-francis.inist.fr/vibad/index.php?action=getRecordDetail&idt=7641732 (accessed on 23 February 2021).
73. Zhang, B.; Zhao, Y.; Li, X.; Zhang, P.; Li, L.; Xie, F.; Chen, L. Effects of amylose and phosphate monoester on aggregation structures of heat-moisture treated potato starches. *Carbohydr. Polym.* **2014**, *103*, 228–233. [CrossRef] [PubMed]
74. Liu, H.; Yu, L.; Xie, F.; Chen, L. Gelatinization of cornstarch with different amylose/amylopectin content. *Carbohydr. Polym.* **2006**, *65*, 357–363. [CrossRef]

75. Stevnebø, A.; Sahlström, S.; Svihus, B. Starch structure and degree of starch hydrolysis of small and large starch granules from barley varieties with varying amylose content. *Anim. Feed Sci. Tech.* **2006**, *130*, 23–38. [CrossRef]
76. Leeman, M.A.; Karlsson, M.E.; Eliasson, A.-C.; Björck, I.M.E. Resistant starch formation in temperature treated potato starches varying in amylose/amylopectin ratio. *Carbohydr. Polym.* **2006**, *65*, 306–313. [CrossRef]
77. Deshpande, S.S.; Cheryan, M. Effects of Phytic Acid, Divalent Cations, and Their Interactions on Amylase Activity. *J. Food Sci. J. Food Sci.* **1984**, *49*, 516–519. [CrossRef]
78. Ralet, M.-C.; Thibault, J.-F.; Della Valle, G. Influence of extrusion-cooking on the physico-chemical properties of wheat bran. *J. Cereal Sci.* **1990**, *11*, 249–259. [CrossRef]
79. Bader Ul Ain, H.; Saeed, F.; Ahmed, A.; Asif Khan, M.; Niaz, B.; Tufail, T. Improving the physicochemical properties of partially enhanced soluble dietary fiber through innovative techniques: A coherent review. *J. Food Process. Preserv.* **2019**, *43*, e13917. [CrossRef]

MDPI

Article

Screening of Twelve Pea (*Pisum sativum* L.) Cultivars and Their Isolates Focusing on the Protein Characterization, Functionality, and Sensory Profiles

Verónica García Arteaga [1,2,*], Sonja Kraus [1], Michael Schott [1], Isabel Muranyi [1], Ute Schweiggert-Weisz [1,3] and Peter Eisner [1,4,5]

1 Fraunhofer Institute for Process Engineering and Packaging IVV, 85354 Freising, Germany; sonni.kraus@gmail.com (S.K.); michael.schott@ivv.fraunhofer.de (M.S.); isabel.muranyi@ivv.fraunhofer.de (I.M.); uweisz@uni-bonn.de (U.S.-W.); peter.eisner@ivv.fraunhofer.de (P.E.)
2 Center of Life and Food Sciences Weihenstephan, Technical University of Munich, 85354 Freising, Germany
3 Institute for Nutritional and Food Sciences, University of Bonn, 53012 Bonn, Germany
4 ZIEL—Institute for Food & Health, Technical University of Munich, 85354 Freising, Germany
5 School of Technology and Engineering, Steinbeis-Hochschule, 12489 Berlin, Germany
* Correspondence: veronica.garcia.arteaga@ivv.fraunhofer.de; Tel.: +49-8161-491-465

Citation: García Arteaga, V.; Kraus, S.; Schott, M.; Muranyi, I.; Schweiggert-Weisz, U.; Eisner, P. Screening of Twelve Pea (*Pisum sativum* L.) Cultivars and Their Isolates Focusing on the Protein Characterization, Functionality, and Sensory Profiles. *Foods* **2021**, *10*, 758. https://doi.org/10.3390/foods10040758

Academic Editors: Christine Scaman and Joana S. Amaral

Received: 4 March 2021
Accepted: 30 March 2021
Published: 2 April 2021

Abstract: Pea protein concentrates and isolates are important raw materials for the production of plant-based food products. To select suitable peas (*Pisum sativum* L.) for protein extraction for further use as food ingredients, twelve different cultivars were subjected to isoelectric precipitation and spray drying. Both the dehulled pea flours and protein isolates were characterized regarding their chemical composition and the isolates were analyzed for their functional properties, sensory profiles, and molecular weight distributions. Orchestra, Florida, Dolores, and RLPY cultivars showed the highest protein yields. The electrophoretic profiles were similar, indicating the presence of all main pea allergens in all isolates. The colors of the isolates were significantly different regarding lightness (L*) and red-green (a*) components. The largest particle size was shown by the isolate from Florida cultivar, whereas the lowest was from the RLPY isolate. At pH 7, protein solubility ranged from 40% to 62% and the emulsifying capacity ranged from 600 to 835 mL g^{-1}. The principal component analysis revealed similarities among certain pea cultivars regarding their physicochemical and functional properties. The sensory profile of the individual isolates was rather similar, with an exception of the *pea-like* and *bitter* attributes, which were significantly different among the isolates.

Keywords: pea (*Pisum sativum* L.); spray-dry; functional properties; sensory profile; protein characterization; pea allergens

1. Introduction

Peas (*Pisum sativum* L.) were domesticated around 10,000 years ago. Over the years, evolution and breeding has influenced the number of pea cultivars found today. In Europe, according to the Food and Agriculture Organization database [1], France and Germany were the biggest dry pea seed producers in 2019. The differences among cultivars depend on their cultivated status (wild or cultivated), geographical origin, and usage (fresh or dry) [2]. The study of different cultivars, their breeding, and their inclusion in the genome database is a continuous process [3]. From an agronomic point of view, cultivation factors such as maximum yield security, plant stability, seed percentage, and protein yield are the most important characteristics considered for pea cultivation; however, for industrial food production, factors such as protein content, functionality, taste, and color are also considered [4]. Peas contain high amounts of protein at around 20–35%, low amounts of fat at around 0.5–4.0%, and high amounts of starch at around 30–48% [5–7]. Previous studies have investigated the differences in pea cultivar compositions and have found environmental and genotypic variations as the main factors for the described data discrepancies.

The aroma of the pea seeds also changes significantly depending on the cultivar, harvest year, and processing conditions [8,9].

Vegetarian or vegan diets might lead to protein deficiencies, making peas an interesting protein source for plant-based food products [10]. According to the Global Market Insights report [11], the pea protein market is estimated to grow by 12% compound annual growth rate (CAGR) by 2026. The main proteins in peas correspond to storage proteins. These are divided into globulins and albumins, corresponding to 55–80% and 18–25%, respectively, depending on genetic and environmental factors [6,12,13]. Similar to other legumes, the major globulins in peas are divided into 7S vicilin–convicilin and 11S legumin fractions [14]. The molecular structures and weight distributions are different among these proteins. Legumin is a hexamer with major polypeptide subunits of ~40 and ~20 kDa, which can be bound by disulfide bonds. Vicilin is a trimer (each subunit ~50 kDa) lacking cysteine residues that can undergo post-translational proteolysis, resulting in different fractions. Convicilin is a trimer (~70 kDa) without any translational modification [15,16].

Pea proteins are used as concentrates (40–90% protein) and isolates (>90% protein) in the food industry; however, the extraction of pea protein isolate (PPIs) at laboratory and pilot scales has shown protein contents of around 80–90% [17–19]. These studies have found that depending on the cultivar and the extraction method, the protein solubility and emulsifying and foaming capacity were significantly affected; however, Stone and Karalash [17] concluded that overall, the extraction method has a greater influence than the cultivar. The PPIs investigated in the above-mentioned studies showed higher functionality than commercial isolates. This could be due to a deviating production or drying process. Industrial protein ingredient suppliers usually use spray drying, whereas lyophilization is mainly used for scientific purposes at the laboratory scale. Spray drying might affect the aroma and protein structure, and thus the protein profile, particle size, and functionality [9,20]. Moreover, most authors have investigated cultivars available in their countries. In Germany, the cultivars Astronaute and Salamanca are mainly used because of their high seed yields [21]; however, to our knowledge, only protein preparations of the latter cultivar have been characterized scientifically [22]. A broader screening of European pea cultivars would increase the ability to select a cultivar that fulfills specific product needs.

Another reason for the high popularity of peas as raw materials for protein isolation is that unlike soy, pea proteins do not need to be declared as allergens in Europe. However, two major allergens, namely convicilin (Pis s 2) and vicilin (Pis s 1), have been identified [23]. Pis s 2 corresponds to a 62–67 kDa fraction, whereas Pis s 1 corresponds to 47–50 kDa (mature vicilin-$\alpha\beta\gamma$) and 32 kDa (vicilin-$\alpha\beta$) fractions. These allergens could potentially promote cross-reactions with other legume allergens; thus, recent studies suggest their inclusion in the allergen declaration list [24,25]. The allergenic potential might vary within and among cultivars, as they have shown significant proteomic variations of the same pea cultivar harvested over three consecutive years [16].

The present study aimed to investigate pea cultivars grown in Germany and France, regarding chemical compositions of their flours and isolates, as well as the protein yields, functional properties, aroma profiles, and molecular weight distribution of the PPIs. Among the data assessed, this study aimed to identify PPIs of cultivars showing similar chemical, functional, and sensory properties in order to use them in combination or interchangeably in the food industry, without having significant effects on the final product quality.

2. Materials and Methods

2.1. Materials

The different field pea seeds (*Pisum sativum* L.) were kindly provided by Norddeutsche Pflanzenzucht Hans-Georg-Lembke KG (Holtsee, Germany) and are shown in Table 1. The Broad Range™ Unstained Standard, 4–20% Criterion™ TGX Stain-Free™ Precast Gels, and Coomassie blue R-250 were purchased from Bio-Rad Laboratories GmbH (Feldkirchen, Germany). Sodium dihydrogen phosphate, sodium dodecyl sulfate, and sodium monohydrogen phosphate were purchased from Sigma-Aldrich Chemie GmbH (Munich, Germany). All chemicals used in this study were of analytical grade.

Table 1. List of pea cultivars investigated in this study.

Cultivar	Harvest Year	Place of Cultivation	Cotyledon Color	Admitted in
Navarro	2018	Malchow/Mecklenburg-Vorpommern	yellow	Germany
Dolores	2015	Oderaue/Mecklenburg-Vorpommern	yellow	Germany
Greenwich	2018	Hohenlieth/Schleswig-Holstein	green	Great Britain
Bluetime	2018	Hohenlieth/Schleswig-Holstein	green	Great Britain
Ostinato	2018	Rodez/France	yellow	France
Kalifa	2017	Hohenlieth/Schleswig-Holstein	yellow	Breeding line
Salamanca	2018	Malchow/Mecklenburg-Vorpommern	yellow	Germany, Czech Republic, etc.
Florida	2015	Dreveskirchen/Mecklenburg-Vorpommern	yellow	Germany
RLPY141091	2018	Rodez/France	yellow	Germany
Orchestra	2018	Rodez/France	yellow	France, Germany
Astronaute	2018	Groß Kiesow/Mecklenburg-Vorpommern	yellow	France, Germany, etc.
Croft	2018	Hohenlieth/Schleswig-Holstein	green	Great Britain

2.2. *Production of Pea Flour*

Peas were dehulled and split using an underflow peeler (Streckel and Schrader KG, Hamburg, Germany). The kernels were separated using a zig-zag airlift system and milled with a pilot plant impact mill with 0.5 mm sieve insertion (Alpine Hoakawa AG, Augsburg, Germany).

2.3. *Production of Pea Protein Isolate*

The isolation of pea protein was performed according to Tian and Kyle [26] following an alkaline extraction with isoelectric precipitation (AE-IEP) with some changes. An aqueous alkaline extract of the pea flour was prepared in deionized (DI) water at a ratio of 1:5 (w/w) at pH 8.0 using 3.0 mol/L NaOH, which was stirred for 60 min. The protein extract was sieved (0.8 mm) after centrifugation at 8000$\times$ g for 20 min at 15 °C (8K, Sigma Laborzentrifugen GmbH, Osterode am Harz, Germany). For isoelectric precipitation, the protein extract was adjusted to pH 4.5 using 3.0 mol/L HCl and left overnight at 4 °C. The precipitated proteins were separated by centrifugation at 8000$\times$ g for 20 min at 15 °C and the protein isolate was dispersed in DI water to a dry matter content of 8%. After neutralization to pH 7.0, the isolate dispersion was homogenized at 11,000 rpm for 2 min using an Ultraturrax (IKA®-Werke GmbH and Co. KG, Staufen, Germany) prior to spray drying. The spray drying was performed using a Mini Spray Dryer B-191 (BUCHI Labortechnik GmbH, Essen, Germany) at inlet and outlet temperatures of 180 °C and 80 °C, respectively, as well as with a 95% aspirator output. The spray-dried isolates were used for further analysis. The protein yield was calculated as grams of protein per kilogram of seeds. Due to the limited amounts of pea seeds, the protein extractions and spray drying were performed once. We assumed that the protein extraction and yield values are representative of the process, as other studies have shown low standard deviations in their own extractions [17,18].

2.4. *Chemical Composition*

The analysis of the chemical compositions of the pea flours and PPIs included determination of the dry matter, ash, protein, starch, and fat contents.

Dry matter and ash contents were determined using thermogravimetric methods (TGA 701, Leco Instruments, Mönchengladbach, Germany). The protein content was determined according to the Dumas combustion method (TruMac N, Leco Instruments, Mönchengladbach, Germany) using the average nitrogen-to-protein conversion factor of N × 6.25. All analyses were performed in duplicate and in accordance with the Association of Official Analytical Collaboration (AOAC) Official Methods [27,28].

The starch content was determined in duplicate using a Starch UV-Test Kit according to the manufacturer's instructions (R-Biopharm AG, Darmstadt, Germany). The fat content was determined according to the Caviezel method [29] with some modifications. In extraction vessels, 2–3 g of the sample was mixed with 1.5 g potassium hydroxide, 5 mL stock solution, and 40 mL 1-butanol. After separation of the derivatized fatty acids by gas chromatography (GC 7890A, Agilent Technologies Germany GmbH & Co. KG, Waldbronn, Germany), the total fat content was determined by summing up all detected methyl esters in relation to an internal standard. Mazola corn germ oil served as the reference. The results are given in fat%, calculated as methyl ester.

2.5. Molecular Weight Distribution Using Sodium Dodecyl Sulfate–Polyacrylamide Gel Electrophoresis (SDS-PAGE)

The molecular weight distribution was analyzed using SDS-PAGE under non-reducing and reducing conditions according to the method used by Laemmli [30], with slight modifications. Briefly, 5 µg/µL protein solution (based on dry matter) was prepared in 1× treatment buffer (50% (*v*/*v*) 2× Tris-HCl treatment buffer, 50% (*v*/*v*) phosphate buffer (pH 7)). The 2× treatment buffer was prepared using 0.125M from the 4× stacking gel buffer (0.5M Tris, adjusted with HCl to pH 6.8), 4% from 10% SDS, 20% glycerol, and 0.02% Bromophenol Blue, while for reduction conditions 0.2M dithiothreitol was added. The samples were heated (95 °C, 5 min) prior to centrifugation at 12,045× *g* for 3 min (MiniSpin, Eppendorf AG, Hamburg, Germany). The supernatants were mixed 1:10 (*v*/*v*) with 1× treatment buffer, from which 3 µL was added into the gel pocket of the Bio-Rad 4–20% Criterion™ TGX Stain-Free™ Precast Gels. The Broad Range™ Unstained Protein Standard was used as the molecular weight marker. The running time was 30 min, followed by staining using Coomassie Brilliant Blue R-250. Finally, gel images were obtained using an EZ Imager (Gel Doc™ EZ Imager—Bio-Rad). Protein bands and their intensities were calculated using Image Lab Software. SDS-PAGE was performed in duplicate, with each sample being prepared two times independently.

2.6. Color

The colors of the protein isolates were measured using the Digi Eye system (VeriVide Limited, Leicester, UK) and a Nikon D90 camera (Nikon Metrology GmbH, Düsseldorf, Germany). The International Commission on Illumination (CIE) L*a*b* method was used to measure the parameters lightness (L*), green-red (a*), and blue-yellow (b*). The white color from the calibration board was used as the white reference for comparison among samples. The total color difference (ΔE^*_{ab}) compared to the white reference board was calculated according to the CIE76 formula (Equation (1)). The color determination was performed in triplicate.

$$\Delta E^*_{ab} = \sqrt{\left(L^*_2 - L^*_1\right)^2 + \left(a^*_2 - a^*_1\right)^2 + \left(b^*_2 - b^*_1\right)^2} \quad (1)$$

2.7. Particle Size

The particle size distribution of all pea protein isolates was determined using a Master-Sizer S Long Bed Version 2.19 equipped with a QS Small Volume Sample Dispersing Unit DIF2021 (Malvern Panalytical Ltd., Malvern, UK). The sample was dispersed in 1-Butanol for 2 min at 3000 rpm before measurements. After another minute, a second measurement was conducted. The measuring range was set at 300 RF 0.05–900 µm. The particle size was based on Mie theory with a refractive index of 1.33, using an index of 0.1 for dispersion media and 1.56 for the dispersed phase, with an imaginary proportion of 0.1.

2.8. Functional Properties

All analyses of functional properties were performed in duplicate.

2.8.1. Protein Solubility

The protein solubility measurements at pH 4.5 and 7.0, respectively, were performed according to Morr and German [31]. The soluble protein content was determined photometrically at 550 nm following the Biuret method [32] using bovine serum albumin (BSA) as the standard for calibration.

2.8.2. Foaming Capacity

The foaming capacity was analyzed at pH 4.5 and 7.0 according to Phillips and Haque [33] using a whipping machine (Hobart N50, Hobart GmbH, Offenburg, Germany). Briefly, 5% (*w*/*v*) dispersions were whipped (580 rpm) for 8 min and the foaming capacities were determined as the relation between the initial and final volume.

2.8.3. Emulsifying Capacity

The emulsifying capacity was determined according to Wang and Johnson [34] and García Arteaga, et al. [35] at pH 4.5 and 7.0. Briefly, 10 mL min^{-1} oil was added to a dispersion (1% *w*/*w*) in a 1 L reactor equipped with an Ultra-Turrax instrument and a conductivity meter. The volume of added oil was used to calculate the emulsifying capacity (mL oil/g sample).

2.9. Sensory Analysis

2.9.1. Sample Preparation

A 2% sample solution (1.7% protein, *w*/*w*) was prepared with tap water for each PPI. The respective samples were adjusted to pH 7.0 with 1 mol/L NaOH and coded using three-digit random numbers.

2.9.2. Sample Evaluation

The sensory evaluation was conducted according to DIN 10967-1-1999 and as described by García Arteaga, et al. [35]. Briefly, a trained panel evaluated attributes regarding retronasal aromas and tastes of the different PPIs. From each sample solution, 20 mL was presented at room temperature in a glass cup and in random order. The sensory evaluation was split into two evaluation sessions. In the evaluation sessions, each panelist evaluated six and seven samples, respectively. The panelists assessed the samples according to the following attributes: *fatty* (2-nonenal); *green* (hexanal); *earthy* (geosmin); *roasty* (2-acetylpyrazine); *pea-like* (2-isopropyl-3-methoxypyrazine); *metallic* ((trans)-4,5-Epoxy-(E)-decenal); *malty*; *nutty* (2,5-dimethylpyrazine). Additionally, panelists assessed the samples according to tastes such as *bitter*, *sweet*, *salty*, *astringent*, *mouth-coating*, and overall intensity. The intensities were scored from 0 (not perceivable) to 10 (very intense).

2.10. Principal Component Analysis

A principal component analysis (PCA) is a multivariate statistical data analysis tool used to simplify the variability of data with a reduced number of dependent variables. A PCA (correlation matrix) was used to evaluate the similarities among isolates regarding their protein content, fat content, color, particle size, and functional properties. A covariance PCA was used to evaluate the aroma and taste. The PCA plots were performed using the software OriginPro 2018b.

2.11. Statistical Analysis

Protein extractions were performed once and the resulting isolates were used for further analyses. Due to the low protein yields, all analyses were performed in duplicate, unless stated otherwise, and the results are expressed as mean values ± standard deviations. Non-parametric statistical analyses were performed due to the low number of replicates. The Kruskal–Wallis test was used to determine statistical differences among the cultivars. Dunn's test with Bonferroni correction for p-values was used as a test for multiple comparisons. The results of the sensory analysis were analyzed using one-way ANOVA followed by Tukey's post hoc test. A Kendall correlation coefficient was used to determine correlations between physicochemical, functional, and sensory properties. All statistical analyses were performed using OriginPro 2018b and were considered statistically significant at $p < 0.05$. The raw data are available as Mendeley Data [36].

3. Results and Discussion

3.1. Chemical Composition and Protein Yield

Table 2 shows the chemical compositions of the flours and PPIs, as well as the protein yields after spray drying.

3.1.1. Pea Flours

The protein contents of the dehulled pea flours ranged from 21.3% to 27.2%, similar to values obtained by Barac and Cabrilo [18] and Nikolopoulou and Grigorakis [6]. The flour from RLPY cultivar showed the highest protein content (27.2%), whereas the flour from Greenwich had the lowest protein content at 21.3%. The ash and fat contents ranged from 2.5% to 3.6% and from 1.9% to 2.5%, respectively. The flour from the Florida cultivar showed the highest fat content of 2.7%, whereas the flours of Dolores, Ostinato, Kalifa, RLPY, and Orchestra cultivars showed the lowest amounts at 1.9%. The flour from Navarro had the highest starch content, while RLPY had the lowest. The protein and starch contents obtained in this study were within the ranges of different cultivars investigated in other studies [22,37].

3.1.2. Pea Protein Isolates

The protein contents of the pea protein isolates (PPIs) ranged from 83.5% to 90.3%. The PPI obtained from the RLPY cultivar showed the highest protein content, while the one from Navarro showed the lowest. The protein contents were in the same range as in other studies [17,18]; however, other studies obtained higher protein yields (62–89%), probably attributed to the drying technique, as high losses are common during spray drying [38]. It is worth mentioning that protein isolation at industrial scale might result in higher yields when the drying kinetics are correctly determined [39,40]. The highest protein yield was 62.2 g protein kg^{-1} $seed^{-1}$ obtained from the Orchestra cultivar, followed by Florida with 59.2 g kg^{-1}. The lowest protein yields were obtained from Navarro and Greenwich cultivars at 33.8 g kg^{-1} and 34.8 g kg^{-1}, respectively. The ash contents of the PPIs varied from 5.3% to 8.5%, probably due to formation of salts (NaCl) after adjusting the pH during the different process steps. The fat contents ranged from 4.7% to 9.0%, with the Greenwich isolate having the highest fat content and Dolores isolate the lowest. The PPIs without a de-fatting step had higher lipid contents, probably due to the protein–lipid interactions during the extraction; Gao and Shen [41] showed that PPIs extracted after AE-IEP had predominantly hydrophobic β-sheets in their protein structures that could promote these interactions [42]. Furthermore, the increase in fat content might promote lipid–protein interactions in the isolates, which may lead to a higher hydrophobic character of the complexes, resulting in lower protein solubility. Their interaction may also reduce the availability of lipophilic groups, limiting the absorption of fat [43]. The color, aroma, and functionality might be also affected by the fat content, especially after lipid oxidation by lipoxygenase [44].

Table 2. Chemical composition and protein yield of dehulled flour and protein isolates produced from different pea cultivars.

Cultivar	Dry Matter	Protein *	Ash 550 *	Fat *	Starch *	Protein Yield **
	[%]	[%]	[%]	[%]	[%]	[g kg^{-1}]
Dehulled Flour						
Navarro	89.6 ± 0.0	22.1 ± 0.1	2.9 ± 0.0	2.3 ± 0.0	52.6 ± 0.2	-
Dolores	91.4 ± 0.0	26.5 ± 0.2	3.0 ± 0.0	1.9 ± 0.0	44.3 ± 2.5	-
Greenwich	90.9 ± 0.0	21.3 ± 0.2	2.7 ± 0.0	2.5 ± 0.0	48.2 ± 0.3	-
Bluetime	91.5 ± 0.1	22.4 ± 0.4	2.9 ± 0.0	2.2 ± 0.0	40.3 ± 0.5	-
Ostinato	91.2 ± 0.0	25.0 ± 0.2	3.6 ± 0.1	1.9 ± 0.0	47.6 ± 0.5	-
Kalifa	91.4 ± 0.1	24.2 ± 0.1	3.0 ± 0.0	1.9 ± 0.0	46.6 ± 0.1	-
Salamanca	90.8 ± 0.0	22.4 ± 0.1	2.8 ± 0.0	2.0 ± 0.0	49.2 ± 3.3	-
Florida	91.2 ± 0.1	24.8 ± 0.1	2.9 ± 0.0	2.7 ± 0.2	45.0 ± 4.6	-
RLPY 141091	91.3 ± 0.1	27.2 ± 0.0	2.8 ± 0.0	1.9 ± 0.0	32.5 ± 0.8	-
Orchestra	92.2 ± 0.1	26.3 ± 0.2	3.5 ± 0.2	1.9 ± 0.0	35.8 ± 0.3	-
Astronaute	91.2 ± 0.0	22.0 ± 0.0	2.5 ± 0.1	2.0 ± 0.1	45.3 ± 1.1	-
Croft	91.8 ± 0.1	22.5 ± 0.1	2.6 ± 0.0	2.1 ± 0.0	48.0 ± 2.2	-
Protein Isolate						
Navarro	93.0 ± 0.0	83.5 ± 0.4	5.3 ± 0.3	5.9 ± 0.0	-	33.8
Dolores	93.5 ± 0.1	89.5 ± 0.2	5.4 ± 0.1	4.7 ± 0.1	-	54.4
Greenwich	93.8 ± 0.0	83.6 ± 0.4	6.0 ± 0.6	9.0 ± 0.2	-	34.8
Bluetime	94.4 ± 0.0	84.1 ± 0.0	6.4 ± 0.4	8.4 ± 0.3	-	42.2
Ostinato	94.1 ± 0.0	86.0 ± 0.5	7.6 ± 0.4	7.1 ± 0.4	-	38.6
Kalifa	93.0 ± 0.0	86.9 ± 0.9	5.9 ± 0.1	7.0 ± 0.5	-	46.2
Salamanca	93.7 ± 0.6	85.0 ± 0.3	6.1 ± 1.0	8.7 ± 0.6	-	42.2
Florida	92.5 ± 0.0	87.4 ± 1.1	5.6 ± 0.1	7.4 ± 0.7	-	59.2
RLPY 141091	93.4 ± 0.0	90.3 ± 0.0	8.5 ± 0.7	7.3 ± 0.8	-	53.6
Orchestra	92.8 ± 0.3	87.1 ± 0.1	6.7 ± 1.1	6.2 ± 0.9	-	62.2
Astronaute	96.0 ± 0.2	86.4 ± 0.1	5.4 ± 0.1	7.8 ± 0.1	-	42.1
Croft	92.5 ± 0.1	86.7 ± 0.6	6.2 ± 0.1	7.8 ± 0.1	-	47.3

Results are expressed as means ± standard deviations ($n = 2$). No significant differences were found among cultivars within the same column (Dunn's test with Bonferroni correction, $p < 0.05$). Note: * based on dry matter; ** based on protein content (g of protein/kg of seeds).

3.2. Molecular Weight Distribution

Gel electrophoresis was performed under non-reducing and reducing conditions to reveal differences within the protein composition of the isolates from the different cultivars (Figure 1). The protein fractions ranged from 93 to 6.5 kDa. Three major fractions were identified in both conditions. Under non-reducing conditions, fractions around ~65 kDa, ~53 kDa, and ~45 kDa were most prominent, while under reducing conditions, 53 kDa proteins were absent and the intensity of the ~39 kDa fraction increased. Bands around 86 and 91 kDa may have been due to convicilin precursors and lipoxygenase (LOX), respectively [16,45]. The most visible difference between non-reducing and reducing conditions was observed for all protein isolates around the 50–56 kDa region, which might correspond to legumin [16,18]. Legumin consists of two polypeptides, one acid (Leg α) and one basic (Leg β) subunit, connected via disulfide bonding. These subunits were found at around 37–40 kDa for Leg α and 19–22 kDa for Leg β, with higher intensities under reducing conditions.

Among the different protein fractions, the allergens Pis s2 and Pis s1 were investigated in detail. Table 3 shows the protein band intensities for each of the allergen fractions. These allergens lack cysteine residues, hindering the formation of disulfide bonds [14]. For this reason, the allergen protein fractions were expected to appear under both conditions.

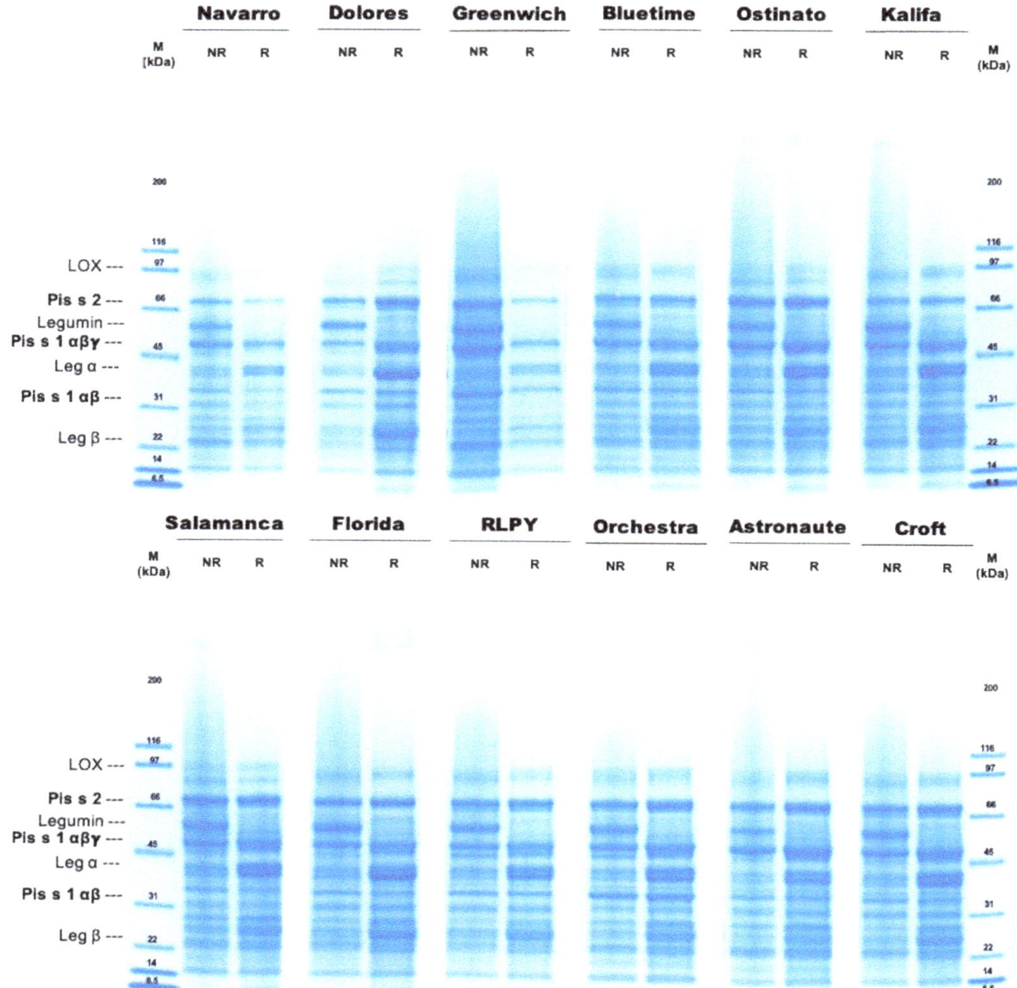

Figure 1. Molecular weight distribution of pea protein isolates from different cultivars, as determined by SDS-PAGE under non-reducing (NR) and reducing (R) conditions. Pis s2, Pis s1 $\alpha\beta\gamma$, and Pis s2 $\alpha\beta$ correspond to the allergen fractions from convicilin, mature vicilin- $\alpha\beta\gamma$, and vicilin-$\alpha\beta$, respectively. M: molecular weight standard indicated in kilodalton (kDa).

Convicilin Pis s2. The average molecular weight of the Pis s2 fraction was around 65 kDa, and values were not significantly different among the isolates. The isolate from Orchestra showed the strongest intensities under both conditions. In contrast, the Kalifa isolate and the Navarro isolate showed the lowest intensities under non-reducing and reducing conditions, respectively. The intensity of this protein fraction increased slightly under reducing conditions for all isolates, except for the Navarro and Greenwich isolates, where the intensity of the bands was slightly lower.

Vicilin Pis s1. The mature allergen fraction (vicilin-$\alpha\beta\gamma$) was around 45 kDa under both conditions and was not significantly different among the isolates. Under non-reducing conditions, the Navarro isolate showed the strongest intensity. On the other hand, the RLPY and Florida isolates showed the lowest intensities. Under reducing conditions, vicilin-αβγ from Astronaute isolate showed the highest intensity, whereas the one from the Greenwich isolate showed the lowest. Vicilin-αβγ can go through post-translational cleavage, resulting in different fractions [46]. From these proteolytic fragments, vicilin-αβ (~32 kDa) was shown to bear a high allergenic potential [23]. Besides the Orchestra isolate, all isolates showed lower intensities of the vicilin-αβ fractions compared to the mature fraction and were not significantly different.

Table 3. Protein band intensities (Int) of globular protein allergens of pea protein isolates, namely convicilin (Pis s2), vicilin αβγ (Pis s1), and vicilin-αβ (Pis s1) from different cultivars as determined by sodium dodecyl sulfate–polyacrylamide gel electrophoresis.

Cultivar	**Protein Band Intensity [Int]**					
	Convicilin Pis s2		**Vicilin αβγ Pis s1**		**Vicilin αβ Pis s1**	
	NR	**R**	**NR**	**R**	**NR**	**R**
Navarro	185 ± 16	146 ± 79	308 ± 34	280 ± 73	81 ± 13	95 ± 50
Dolores	214 ± 5	367 ± 35	166 ± 22	282 ± 43	118 ± 5	97 ± 11
Greenwich	219 ± 47	185 ± 57	229 ± 55	256 ± 36	126 ± 16	113 ± 2
Bluetime	241 ± 17	263 ± 49	236 ± 66	311 ± 39	111 ± 7	127 ± 9
Ostinato	253 ± 31	343 ± 32	252 ± 66	361 ± 11	105 ± 19	116 ± 37
Kalifa	141 ± 6	205 ± 36	285 ± 47	392 ± 46	122 ± 8	153 ± 7
Salamanca	280 ± 2	363 ± 5	233 ± 51	350 ± 20	89 ± 3	88 ± 16
Florida	218 ± 16	283 ± 51	196 ± 88	330 ± 60	94 ± 1	72 ± 6
RLPY 141091	294 ± 13	379 ± 63	149 ± 25	285 ± 78	117 ± 8	106 ± 0
Orchestra	302 ± 50	421 ± 32	212 ± 78	398 ± 19	228 ± 38	180 ± 5
Astronaute	251 ± 34	365 ± 44	293 ± 72	411 ± 20	118 ± 28	99 ± 8
Croft	261 ± 55	372 ± 79	235 ± 79	353 ± 29	93 ± 7	79 ± 20

Results are expressed as means ± standard deviations (n = 2). No significant differences were found among cultivars within the same column (Dunn's test with Bonferroni correction, $p < 0.05$). NR: non reducing conditions; R: reducing conditions.

The intensities of the protein and allergen fractions can differ among legume cultivars, while the intensity of the allergen fractions specifically can give an indication of the allergenic potential [14,18,47]. Overall, the Orchestra isolate showed slightly stronger intensities for the potential allergen fractions compared to the other isolates, whereas the isolate from Navarro had the lowest intensities for these fractions. However, the allergen fraction intensities were not significantly different among isolates. It is known that the globulin-to-albumin and legumin-to-vicilin ratios change throughout seed development [48], which could affect the presence and intensity of potential allergens. Even under the same environment, harvesting, and storage conditions, the variation among proteins in pea cultivars can be very large [14].

3.3. Color

Table 4 shows the color values of the samples and the white reference. The lightness (L*) levels among isolates were significantly different; the isolate of Orchestra cultivar showed significantly higher lightness (90.6) than the Bluetime (86.8) isolate. The Greenwich, Bluetime, and Croft isolates showed the lowest a* values, which corresponded to their cotyledon green color; however, only the isolate from Croft was significantly different to the isolates from Salamanca and Astronaute cultivars. In contrast, the isolate from the Navarro cultivar showed higher b* values, suggesting a stronger yellow color. The total color difference (ΔE^*_{ab}) allows for quantification of the colors and allows comparison between samples; the lower the ΔE^*_{ab} value, the whiter the isolate is. All ΔE^*_{ab} values ranged between 19.2 and 23.4. According to the lowest difference from the white reference, the isolates from the Dolores and Greenwich cultivars were most white, while the isolate from Navarro cultivar was least white.

Table 4. CIE lab color results from pea protein isolates from different cultivars and a commercial pea protein isolate.

Cultivar	Pea Isolate CIE Color			
	L*	a*	b*	ΔE^*_{ab}
Navarro	89.3 ± 0.4 $_{ab}$	2.8 ± 0.1 $_{ab}$	23.7 ± 0.2 $_{a}$	23.4 ± 0.2 $_{a}$
Dolores	88.5 ± 0.1 $_{ab}$	1.9 ± 0.0 $_{ab}$	19.1 ± 0.2 $_{a}$	19.2 ± 0.1 $_{a}$
Greenwich	88.2 ± 0.1 $_{ab}$	0.6 ± 0.1 $_{ab}$	19.2 ± 0.4 $_{a}$	19.3 ± 0.3 $_{a}$
Bluetime	86.8 ± 0.3 $_{a}$	0.9 ± 0.1 $_{ab}$	20.5 ± 0.7 $_{a}$	21.0 ± 0.7 $_{a}$
Ostinato	89.4 ± 0.3 $_{ab}$	3.2 ± 0.2 $_{ab}$	20.5 ± 0.4 $_{a}$	20.4 ± 0.5 $_{a}$
Kalifa	89.5 ± 0.3 $_{ab}$	2.8 ± 0.0 $_{ab}$	20.5 ± 0.2 $_{a}$	20.3 ± 0.2 $_{a}$
Salamanca	88.3 ± 0.2 $_{ab}$	3.3 ± 0.1 $_{ab}$	21.3 ± 0.3 $_{a}$	21.5 ± 0.3 $_{a}$
Florida	88.6 ± 0.3 $_{ab}$	2.7 ± 0.2 $_{ab}$	20.8 ± 0.2 $_{a}$	20.8 ± 0.3 $_{a}$
RLPY 141091	90.1 ± 0.2 $_{ab}$	3.1 ± 0.1 $_{ab}$	22.0 ± 0.2 $_{a}$	22.0 ± 0.3 $_{a}$
Orchestra	90.6 ± 0.6 $_{b}$	2.6 ± 0.3 $_{ab}$	20.9 ± 0.4 $_{a}$	20.3 ± 0.5 $_{a}$
Astronaute	88.2 ± 0.3 $_{ab}$	3.5 ± 0.1 $_{a}$	22.8 ± 0.3 $_{a}$	22.9 ± 0.3 $_{a}$
Croft	87.3 ± 0.3 $_{ab}$	-0.5 ± 0.0 $_{b}$	19.9 ± 0.2 $_{a}$	20.3 ± 0.3 $_{a}$

Results are expressed as means ± standard deviations (n = 3). Subscripts with different letters indicate significant differences within the same column (Dunn's test with Bonferroni correction, $p < 0.05$). Note: ΔE^*_{ab}: color difference compared to a white reference.

3.4. Particle Size

Spray drying is one of the most common methods for drying protein solutions on an industrial scale. However, protein structures are to known to be affected by spray drying due to applied temperatures, vaporization, and the air–water interface.

These effects can cause protein denaturation and further aggregation of the exposed hydrophobic regions, which can affect the particle size of the dried proteins [20]. The particle size, in turn, is known to affect the physicochemical properties of proteins [49,50]. The particle sizes of the PPIs, described as the average volume weighted mean ($d_{4,3}$), are shown in Table 5. The average $d_{4,3}$ of the cultivar isolates was 11.9 µm. Of all protein isolates, the Florida protein isolate showed the largest $d_{4,3}$ at 18.8 µm, followed by the Dolores and Croft isolates. The isolate from RLPY showed the smallest $d_{4,3}$ at 7.5 µm, followed by Ostinato and Astronaute isolates. The different particle sizes among the investigated cultivar isolates might lead to differences in physicochemical behavior as a result of different particle morphologies [51].

Table 5. Physicochemical and functional properties of pea protein isolates from different pea cultivars.

	Particle Size	Protein Solubility **		Emulsifying Capacity		Foaming Capacity
Cultivar	$d_{4,3}$	pH 4.5	pH7.0	pH 4.5	pH 7.0	pH 4.5
	[µm]	[%]	[%]	[mL g^{-1}]	[mL g^{-1}]	[%]
Navarro	13.19 ± 0.56	10.3 ± 0.2	51.5 ± 0.9	405 ± 1	600 ± 7	805 ± 0
Dolores	15.81 ± 0.06	7.4 ± 0.0	60.8 ± 2.8	340 ± 7	706 ± 14	808 ± 4
Greenwich	12.82 ± 0.19	8.8 ± 1.3	55.4 ± 3.4	396 ± 2	734 ± 7	839 ± 36
Bluetime	9.20 ± 0.49	7.7 ± 0.2	53.8 ± 2.4	365 ± 1	710 ± 8	915 ± 0
Ostinato	7.86 ± 0.02	8.3 ± 1.9	60.4 ± 1.9	385 ± 14	787 ± 32	959 ± 10
Kalifa	13.55 ± 1.53	7.3 ± 0.0	40.0 ± 2.1	354 ± 1	747 ± 3	911 ± 40
Salamanca	10.15 ± 0.40	5.9 ± 0.6	48.6 ± 3.6	378 ± 11	744 ± 2	835 ± 0
Florida	18.84 ± 1.31	0.9 ± 1.3	41.3 ± 7.1	340 ± 7	781 ± 23	884 ± 14
RLPY 141091	7.53 ± 0.01	2.3 ± 0.6	52.6 ± 2.8	359 ± 5	835 ± 7	874 ± 13
Orchestra	11.31 ± 0.21	1.5 ± 0.0	61.8 ± 6.0	366 ± 1	790 ± 6	835 ± 9
Astronaute	7.94 ± 0.29	6.3 ± 0.3	52.4 ± 0.9	381 ± 7	681 ± 23	858 ± 23
Croft	14.66 ± 1.35	0.0 ± 0.0	43.6 ± 5.1	355 ± 0	790 ± 24	861 ± 6

Results are expressed as means ± standard deviations. No significant differences were found among cultivars within the same column (Dunn's test with Bonferroni correction, $p < 0.05$). The particle size was based on Mie's theory (RI1.33). Note: $d_{4,3}$: volume weighted mean; ** based on protein content.

3.5. Functional Properties

High functional properties of PPIs are desired to increase their usage as ingredients in different plant-based food products. Table 5 shows the results of the functional properties.

3.5.1. Protein Solubility

At pH 4.5, the Navarro isolate showed the highest protein solubility at 10.3%, which was different to the isolate from Florida (0.9%), Orchestra (1.5%), and Croft (0.0%) cultivars. On the other hand, at pH 7.0, the Orchestra isolate showed the highest protein solubility (61.8%), followed by the Dolores (60.8%) and Ostinato (60.4%) isolates. Overall, the protein solubility levels at pH 7 were similar among the isolates. Other studies have shown similar solubilities or even values up to 80% at pH 7.0 [17–19]. The protein solubility level is related to extraction and drying methods; for example, isolates obtained after alkaline extraction and isoelectric precipitation have lower solubility than those obtained after salt-induced extraction [17]. Moreover, in contrast to lyophilization used in previous studies, spray drying leads to higher protein denaturation, increasing hydrophobic protein–protein interactions, and thus reducing overall protein solubility [52]. High protein solubility levels are, however, essential for beverage and dairy-alternative applications; treatments such as proteolysis or the addition of L-Arginine and sodium carbonate are known to improve protein solubilities of PPIs [22,35].

3.5.2. Emulsifying Capacity

The isolate from Navarro showed the highest emulsifying capacity at pH 4.5 with 405 mL g^{-1}, while the one from Dolores and Florida showed the lowest. There were significant moderate correlations between the emulsifying capacity at pH 4.5 and both the protein solubility at pH 4.5 ($r = 0.50$) and the protein content ($r = -63$). On the other hand, the emulsifying capacity at pH 7.0 showed a significant positive moderate correlation with the protein content ($r = 0.45$). Thus, at neutral pH, the RLPY isolate showed the highest emulsifying capacity at 835 mL g^{-1} and was highly different from the isolates from Navarro and Astronaute cultivars. Hydrophobic residues are essential to facilitate protein oil interactions [53], however a high number of protein–protein interactions would form aggregates hiding hydrophobic residues, thus hindering the ability to interact with oil. These aggregates might be formed during spray drying, thus increasing particle size. However, no significant correlations were found between the emulsifying capacity and the particle size. Moreover, the vicilin/legumin ratio plays an important role in the formation of emulsions; Barac and Cabrilo [18] showed that the lower the ratio is, the higher the emulsifying capacity of the isolate, especially at neutral pH ranges. Although electrophoretic results showed no significant differences among allergens or overall in the electrophoretic patterns, further quantification of the fractions might be necessary to determine correlations with the functional properties.

3.5.3. Foaming Capacity

At pH 7.0 no foam formation was observed, whereas at pH 4.5 all isolates showed an average foaming capacity of 866%. These results are in contrast to the results of Chao and Aluko [54], who obtained higher foaming capacities the further the pH moved away from the isoelectric point. On the other hand, Gharsallaoui and Cases [55] suggested that close to the isoelectric point (pH 4.5), pea globulins are more surface-active and a reduction in the electrostatic charge of the protein molecules might result in electrostatic repulsion, in turn increasing adsorption. The latter is important for the formation of foam [56] and might explain the foaming capacity at pH 4.5 for the cultivars investigated in this study. Another explanation is that the fat content in the PPIs might have acted as an antifoam agent. In order to destroy a foam film, the hydrophobic particle droplets that emerge from the aqueous phase into the air–water interface are critical [57]. At pH 7, the hydrophobic protein surfaces facilitate the entrance of the fat droplets, leading to defoaming. At pH 4.5, the hydrophobic side chains of the proteins are hidden, hindering the penetration of the fat droplets in the foam films.

A principal component analysis (PCA) was applied to analyze the relationships among the different cultivars and their colors, protein and fat contents, particle sizes, and physicochemical properties. Figure 2A shows a biplot of principal component (PC)1 and PC2 using the standardized scores for the isolates. The first two components of the PCA explained 57.13% of the total variance. The protein content (−0.44) and emulsifying capacity at pH 4.5 (0.51) had the strongest influence on PC1. On the other hand, the fat content (0.58) and foaming capacity (0.54) had the strongest influence on PC2; moreover, on the negative quadrant of the PC2, the particle size showed a strong influence (−0.41).

The isolate from Navarro cultivar scored the highest for PC1 (1.93), opposite to the isolates from Kalifa, RLPY, and Croft. This is in agreement with the emulsifying capacities shown in Section 3.5. Furthermore, the Dolores isolate scored the highest in the PC2 (−2.30), followed by Navarro (−1.51), as they showed lower fat contents among the isolates. Negative moderate correlations were found between the protein content and the protein solubility (pH 4.5) and emulsifying capacity (pH 4.5). On the other hand, the protein content was significantly positive and moderately correlated with the emulsifying capacity at pH 7.0. The particle size showed no significant correlations to other investigated attributes. When replacing a raw material in an existing product, not only are the composition and functionality important, but the color should be also considered, as it can affect the perception of the product by the consumer; for this reason, the ΔE^*_{ab} of the isolates was included in the PCA. However, the ΔE^*_{ab} showed low influence on any of the components.

The PCA shows two clusters plus two outliers. The isolates from RLPY, Croft, Kalifa, Florida, and Orchestra cultivars formed the first cluster; on the opposite side, isolates from Ostinato, Bluetime, Salamanca, Astronaute, and Greenwich cultivars formed the second cluster. These clusters suggest that the physicochemical characteristics are probably more similar and one cultivar could be replaced with another from the same cluster. On the other hand, the isolates from Navarro and Dolores were found to be further away from all other isolates, which might hinder the replacement of these cultivars. Moreover, the isolates should be chosen by considering the requirements of the final products. For example, the RLPY isolate could be used in applications with neutral pH, such as dairy alternatives, as it is plotted as having higher protein content, high emulsifying capacity (pH 7.0), and moderate protein solubility (pH 7.0); however, its application at low pH values is inappropriate due to its lower protein functionality. On the other hand, the Navarro isolate might be better suited in applications with acidic pH values, such as in plant-based mayonnaise.

3.6. Sensory Analysis

A principal component analysis was applied to analyze relationships between samples and retronasal aroma attributes and taste profiles (Figure 2B). PC1 and PC2 represented 66.03% of the total variance; the following values represent the coefficient values (influence) of the attributes and the scores of the isolates from each cultivar.

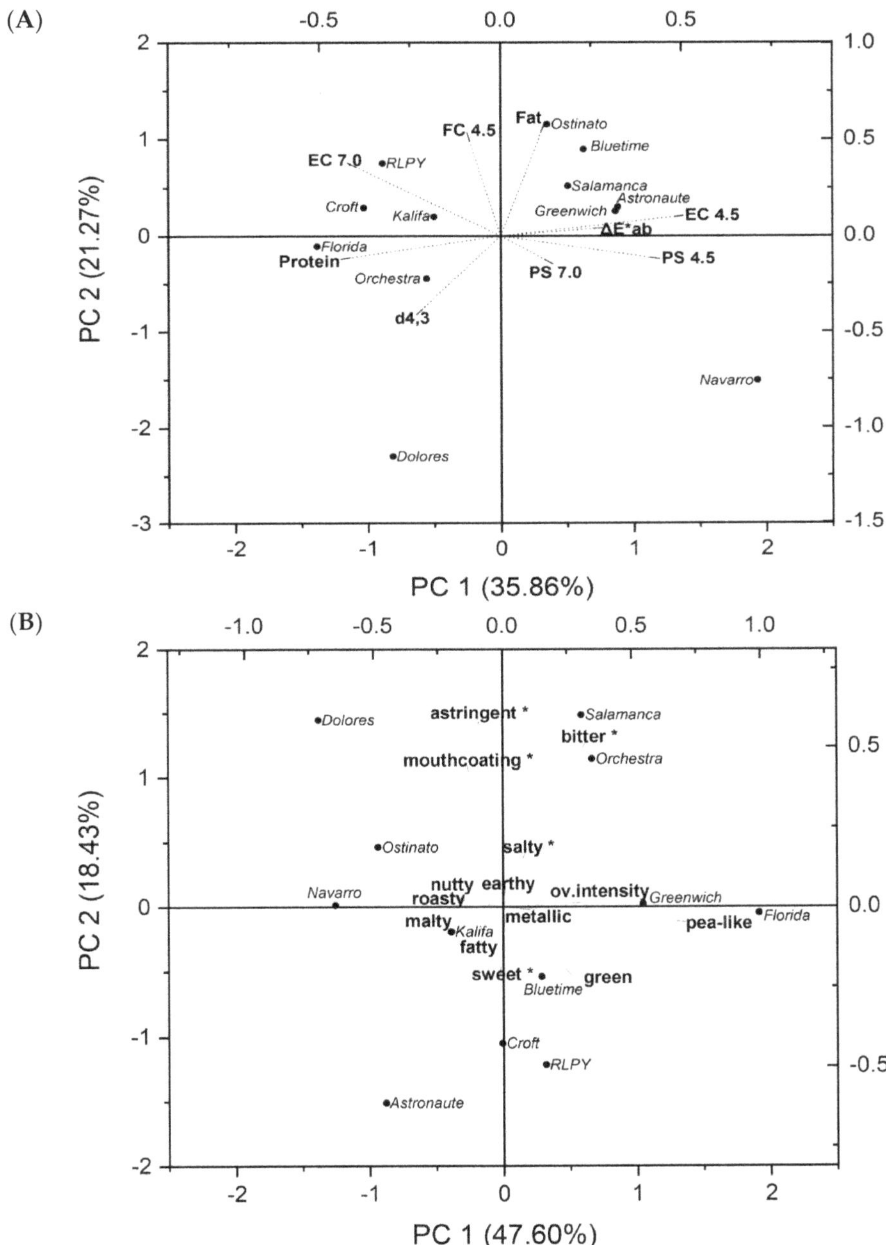

Figure 2. Biplot of (**A**) physicochemical properties and (**B**) sensory profiles of pea protein isolates from different pea cultivars. Attributes with an asterisk (*) refer to taste. PS: protein solubility; EC: emulsifying capacity; FC: foaming capacity; ΔE^*_{ab}: color difference compared to a white reference; $d_{4,3}$: particle size. The numbers represent the pH (7.0 or 4.5) in which the analysis was performed.

Aroma. According to a one-way ANOVA, *pea-like* was the only aroma attribute with a significant difference among isolates and showed the strongest influence on PC1 (0.71), followed by *malty* (−0.30) and *green* (0.27) aromas. For PC2, the *green* aroma showed the strongest influence among all aroma attributes (−0.22). The *metallic*, *earthy*, *roasty*, and *nutty* attributes showed almost no influence on any of the components. The isolates from Dolores (−1.38) and Navarro (−1.25) scored the lowest for PC1, which suggests these isolates were perceived to have the least *pea-like* aroma. In contrast, Florida (1.92) and Greenwich (1.05) isolates scored the highest for this component, indicating a stronger *pea-like* aroma. The *pea-like* aroma is known to be well-perceived because of the low thresholds of 2-isopropyl-3-methoxypyrazine [58], which might explain its strong influence. The PCA showed the isolates from Greenwich, Florida, RLPY, and Croft cultivars as being closer to the *green* attribute, in agreement with the results from the sensory analysis. The *green* aroma originating from hexanal is a characteristic oxidation product of fatty acids, in particular of linoleic and linolenic acids catalyzed by LOX [44]. Higher activity of this enzyme could increase the *green* aroma perception; however, there was only a moderate negative correlation (r = −0.54) between the *green* aroma and the LOX band intensities under reduction conditions. Although it has also been mentioned that green cultivars have higher levels of hexanal [8], there was no significant correlation between the a* color and *green* aroma.

Environmental and genetic conditions might affect the production and degradation rates of the aroma compounds, which would result in higher or lower aroma perceptions [8,59]. Using HC/MS analysis, Azarnia and Boye [8] found that the concentrations of volatile compounds depended on the cultivar, crop year, storage, and processing conditions. Specific processing methods such as enzymatic treatment or fermentation might be useful to reduce some of the characteristic off-flavors of pea isolates. However, other aroma compounds might be generated or enhanced and might further increase or decrease consumer acceptance [4,60,61]. Furthermore, methoxypyrazines are very stable during fermentation due to their chemical nature, and therefore are very difficult to remove or reduce [62]. Therefore, pea cultivars low in *pea-like* aroma, such as from Dolores or Navarro, are recommended to be used for production of PPIs with sensory appeal.

Taste. The *bitter* attribute was the only significant taste attribute according to one-way ANOVA. The *bitter* (0.53) and *astringent* (0.60) tastes had the strongest influence on PC2. As shown in the PCA, the PPIs from Salamanca and Orchestra scored highest for bitter taste. In contrast, the PPI from Astronaute scored lowest for *bitter* taste and was significantly less *bitter* than the PPI from Salamanca. The Dolores isolate scored the highest for *astringent* taste, together with Salamanca and Orchestra isolates. Moreover, *salty* and *sweet* tastes had little influence on either component, which suggests that the intensity of these tastes was lower and similar among the samples. A high *bitter* taste for a PPI might hinder its application in food products; thus, several methods have been investigated to reduce the bitterness of legume protein isolates [61,63].

Overall Intensity. The isolate from Florida cultivar showed the highest overall intensity, whereas the isolate from Dolores showed the lowest intensity. However, the overall intensity levels among the PPIs were not significantly different. The overall intensity was moderately correlated with the *pea-like* (r = 0.66), *green* (r = 0.43), and *malty* (r = −0.47) aroma, which suggests that these compounds were characteristic of the isolates, as mentioned previously in the aroma section.

4. Conclusions

Peas are a valuable source of protein and are increasingly used in plant-based products; however, due to the large number of different cultivars, most of them have not been characterized regarding their chemical composition, functional properties, and sensory profiles. In this study, all these aspects were investigated for 12 cultivars grown in Germany and France.

Our study shows that the chemical composition of flour and isolates from the cultivars are slightly different. The main allergen fractions were present in all the PPI and showed no significant differences. The PCA showed two cluster of cultivars regarding the physicochemical and functional characteristics; however, these clusters were not found in the sensory profile PCA. This suggests that although some isolates could be substituted interchangeably for the same products with regard to their similar functionalities, the flavor of these products could be affected. However, only the *pea-like* and *bitter* aromas were significantly different among isolates. The cultivars Salamanca and Astronaute are the most used cultivars in Germany; they showed similarities according to the physicochemical-functional PCA cluster; however, Salamanca isolates had a significant higher bitter taste and slightly higher *pea-like* aroma than Astronaute. These differences should be considered for targeted product developments as they might influence the acceptance by consumers. The usage of cultivars such as Navarro and Dolores should be carefully considered, as their isolates are mostly different to the other cultivars investigated in the present study.

The obtained PPI might be used in the food industry, especially under neutral conditions (pH 7.0), except when foaming is required; however, when designing a food product in the acidic range at pH 4.5, the specific selection of a suitable cultivar might be more important. Differences with laboratory and commercial processing of PPI should be considered; although spray drying was used in this study, larger spray-dryers may affect the physicochemical, functional and sensory properties of the isolates. The results of this study highlighted the importance of a tailored selection of cultivars for protein extraction as well as the suitability of pea cultivars for specific food applications.

Author Contributions: V.G.A.: conceptualization, methodology, investigation, formal analysis, writing—original draft, review and editing; S.K.: investigation; M.S.: investigation, review and editing; I.M.: methodology, writing—review and editing, supervision; U.S.-W.: resources, review and editing, supervision; P.E.: resources, review and editing, supervision. All authors have read and agreed to the published version of the manuscript.

Funding: This work was supported by the Fraunhofer Future Foundation, Germany.

Informed Consent Statement: Informed consent was obtained from all subjects involved in the sensory analysis.

Data Availability Statement: Raw data are available as Mendeley Data doi:10.17632/ywpjs6h3jr.1.

Acknowledgments: The authors thank Sigrid Gruppe and Eva Müller for their valuable contribution. We also thank Andrea Strube and Thorsten Tybussek for their constructive criticism of the manuscript. We greatly appreciate the sensory panel of the Fraunhofer Institute for Process Engineering and Packaging, Freising, Germany, for the sensory evaluation.

Conflicts of Interest: The authors declare no conflict of interest.

References

1. FAOSTAT. Crops Production. Available online: http://www.fao.org/faostat/en/#data/QC (accessed on 11 January 2021).
2. Kreplak, J.; Madoui, M.-A.; Cápal, P.; Novák, P.; Labadie, K.; Aubert, G.; Bayer, P.E.; Gali, K.K.; Syme, R.A.; Main, D.; et al. A reference genome for pea provides insight into legume genome evolution. *Nat. Genet.* **2019**, *51*, 1411–1422. [CrossRef]
3. Jain, S.; Kumar, A.; Mamidi, S.; McPhee, K. Genetic Diversity and Population Structure Among Pea (*Pisum sativum* L.) Cultivars as Revealed by Simple Sequence Repeat and Novel Genic Markers. *Mol. Biotechnol.* **2014**, *56*, 925–938. [CrossRef]
4. Tulbek, M.C.; Lam, R.S.H.; Wang, Y.; Asavajaru, P.; Lam, A. Chapter 9—Pea: A Sustainable Vegetable Protein Crop. In *Sustainable Protein Sources*; Nadathur, S.R., Wanasundara, J.P.D., Scanlin, L., Eds.; Academic Press: San Diego, CA, USA, 2017; pp. 145–164. [CrossRef]
5. Vidal-Valverde, C.; Frias, J.; Hernández, A.; Martín-Alvarez, P.J.; Sierra, I.; Rodríguez, C.; Blazquez, I.; Vicente, G. Assessment of nutritional compounds and antinutritional factors in pea (*Pisum sativum*) seeds. *J. Sci. Food Agric.* **2003**, *83*, 298–306. [CrossRef]
6. Nikolopoulou, D.; Grigorakis, K.; Stasini, M.; Alexis, M.N.; Iliadis, K. Differences in chemical composition of field pea (*Pisum sativum*) cultivars: Effects of cultivation area and year. *Food Chem.* **2007**, *103*, 847–852. [CrossRef]
7. Al-Karaki, G.; Ereifej, K. Relationships between Seed Yield and Chemical Composition of Field Peas Grown under Semi-arid Mediterranean Conditions. *J. Agron. Crop Sci.* **2001**, *182*, 279–284. [CrossRef]

8. Azarnia, S.; Boye, J.I.; Warkentin, T.; Malcolmson, L.; Sabik, H.; Bellido, A.S. Volatile flavour profile changes in selected field pea cultivars as affected by crop year and processing. *Food Chem.* **2011**, *124*, 326–335. [CrossRef]
9. Cui, L.; Kimmel, J.; Zhou, L.; Rao, J.; Chen, B. Identification of extraction pH and cultivar associated aromatic compound changes in spray dried pea protein isolate using untargeted and targeted metabolomic approaches. *J. Agric. Food Res.* **2020**, *2*, 100032. [CrossRef]
10. Kniskern, M.A.; Johnston, C.S. Protein dietary reference intakes may be inadequate for vegetarians if low amounts of animal protein are consumed. *Nutrition* **2011**, *27*, 727–730. [CrossRef]
11. Ahuja, K.; Mamtani, K. *Pea Protein Market Share Forcasts 2020–2026*; GMI362; Global Market Insights: Selbyville, DE, USA, 2020; p. 180.
12. Koyoro, H.; Powers, J.R. Functional properties of pea globulin fractions. *Cereal Chem.* **1987**, *64*, 97–101.
13. Gueguen, J.; Barbot, J. Quantitative and qualitative variability of pea (*Pisum sativum* L.) protein composition. *J. Sci. Food Agric.* **1988**, *42*, 209–224. [CrossRef]
14. Tzitzikas, E.N.; Vincken, J.P.; de Groot, J.; Gruppen, H.; Visser, R.G. Genetic variation in pea seed globulin composition. *J. Agric. Food Chem.* **2006**, *54*, 425–433. [CrossRef] [PubMed]
15. Casey, R.; Domoney, C. Pea Globulins. In *Seed Proteins*; Shewry, P.R., Casey, R., Eds.; Springer: Dordrecht, The Netherlands, 1999; pp. 171–208. [CrossRef]
16. Bourgeois, M.; Jacquin, F.; Savois, V.; Sommerer, N.; Labas, V.; Henry, C.; Burstin, J. Dissecting the proteome of pea mature seeds reveals the phenotypic plasticity of seed protein composition. *Proteomics* **2009**, *9*, 254–271. [CrossRef] [PubMed]
17. Stone, A.K.; Karalash, A.; Tyler, R.T.; Warkentin, T.D.; Nickerson, M.T. Functional attributes of pea protein isolates prepared using different extraction methods and cultivars. *Food Res. Int.* **2015**, *76*, 31–38. [CrossRef]
18. Barac, M.; Cabrilo, S.; Pesic, M.; Stanojevic, S.; Zilic, S.; Macej, O.; Ristic, N. Profile and functional properties of seed proteins from six pea (*Pisum sativum*) genotypes. *Int. J. Mol. Sci.* **2010**, *11*, 4973–4990. [CrossRef] [PubMed]
19. Tanger, C.; Engel, J.; Kulozik, U. Influence of extraction conditions on the conformational alteration of pea protein extracted from pea flour. *Food Hydrocoll.* **2020**, *107*, 105949. [CrossRef]
20. Haque, M.A.; Adhikari, B. Drying and Denaturation of Proteins in Spray Drying Process. In *Handbook of Industrial Drying*, 4th ed.; CRC Press; Taylor and Francis Group: Boca Raton, FL, USA, 2014; Volume 49, pp. 971–984.
21. LfL. Bayerische Landesanstalt für Landwirtschaft. Available online: https://www.lfl.bayern.de/ipz/oelfruechte/066649/index.php (accessed on 11 January 2021).
22. Reinkensmeier, A.; Bußler, S.; Schlüter, O.; Rohn, S.; Rawel, H.M. Characterization of individual proteins in pea protein isolates and air classified samples. *Food Res. Int.* **2015**, *76*, 160–167. [CrossRef]
23. Sanchez-Monge, R.; Lopez-Torrejon, G.; Pascual, C.Y.; Varela, J.; Martin-Esteban, M.; Salcedo, G. Vicilin and convicilin are potential major allergens from pea. *Clin. Exp. Allergy* **2004**, *34*, 1747–1753. [CrossRef] [PubMed]
24. Dreyer, L.; Astier, C.; Dano, D.; Hosotte, M.; Jarlot-Chevaux, S.; Sergeant, P.; Kanny, G. Consommation croissante d'aliments contenant du pois jaune: Un risque d'allergie? *Rev. Fr. Allergol.* **2014**, *54*, 20–26. [CrossRef]
25. Codreanu-Morel, F.; Morisset, M.; Cordebar, V.; Larré, C.; Denery-Papini, S. L'allergie au pois. *Rev. Fr. Allergol.* **2019**, *59*, 162–165. [CrossRef]
26. Tian, S.J.; Kyle, W.S.A.; Small, D.M. Pilot scale isolation of proteins from field peas (*Pisum sativum* L.) for use as food ingredients. *Int. J. Food Sci. Technol.* **1999**, *34*, 33–39. [CrossRef]
27. AOACa. Method 923.03. Ash of flour. In *Official Methods of Analysis of the Association of Official Analytical Chemists (AOAC)*; Horwitz, W.A., Ed.; AOAC International: Gaithersburg, MD, USA, 2003.
28. AOACb. Method 968.06. Protein (crude) in animal feed. In *Official Methods of Analysis of the Association of Official Analytical Chemists (AOAC)*; Horwitz, W.A., Ed.; AOAC International: Gaithersburg, MD, USA, 2003.
29. Gertz, C.; Fiebig, H.-J. Determination of fat content by the Caviezel® method (rapid method). *Eur. J. Lipid Sci. Technol.* **2000**, *102*, 154–158. [CrossRef]
30. Laemmli, U.K. Cleavage of structural proteins during assembly of head of bacteriophage-T4. *Nature* **1970**, *227*, 680–685. [CrossRef]
31. Morr, C.V.; German, B.; Kinsella, J.E.; Regenstein, J.M.; Vanburen, J.P.; Kilara, A.; Lewis, B.A.; Mangino, M.E. A collaborative study to develop a standarized food protein solubility procedure. *J. Food Sci.* **1985**, *50*, 1715–1718. [CrossRef]
32. AACC Method 46-15.01 Crude Protein—5-Minute Biuret Method for Wheat and Other Grains. Available online: https://methods.aaccnet.org/summaries/46-15-01.aspx (accessed on 1 April 2021).
33. Phillips, L.G.; Haque, Z.; Kinsella, J.E. A Method for the Measurement of Foam Formation and Stability. *J. Food Sci.* **1987**, *52*, 1074–1077. [CrossRef]
34. Wang, C.; Johnson, L.A. Functional properties of hydrothermally cooked soy protein products. *J. Am. Oil Chem. Soc.* **2001**, *78*, 189–195. [CrossRef]
35. García Arteaga, V.; Apéstegui Guardia, M.; Muranyi, I.; Eisner, P.; Schweiggert-Weisz, U. Effect of enzymatic hydrolysis on molecular weight distribution, techno-functional properties and sensory perception of pea protein isolates. *Innov. Food Sci. Emerg. Technol.* **2020**, *65*, 102449. [CrossRef]
36. García Arteaga, V.; Kraus, S.; Schott, M.; Muranyi, I.; Schweiggert-Weisz, U.; Eisner, P. Screening of Twelve Pea (*Pisum sativum* L.) Cultivars and its Isolates. *Mendeley Data* **2021**, *V2*. [CrossRef]

37. Shen, S.; Hou, H.; Ding, C.; Bing, D.-J.; Lu, Z.-X.; Navabi, A. Protein content correlates with starch morphology, composition and physicochemical properties in field peas. *Can. J. Plant Sci.* **2016**, *96*, 404–412. [CrossRef]
38. Maury, M.; Murphy, K.; Kumar, S.; Shi, L.; Lee, G. Effects of process variables on the powder yield of spray-dried trehalose on a laboratory spray-dryer. *Eur. J. Pharm. Biopharm.* **2005**, *59*, 565–573. [CrossRef]
39. Sosnik, A.; Seremeta, K.P. Advantages and challenges of the spray-drying technology for the production of pure drug particles and drug-loaded polymeric carriers. *Adv. Colloid Interface Sci.* **2015**, *223*, 40–54. [CrossRef]
40. Zbicinski, I. Modeling and Scaling Up of Industrial Spray Dryers: A Review. *J. Chem. Eng. Jpn.* **2017**, *50*, 757–767. [CrossRef]
41. Gao, Z.; Shen, P.; Lan, Y.; Cui, L.; Ohm, J.-B.; Chen, B.; Rao, J. Effect of alkaline extraction pH on structure properties, solubility, and beany flavor of yellow pea protein isolate. *Food Res. Int.* **2020**, *131*, 109045. [CrossRef]
42. Sternberg, M.J.E.; Thornton, J.M. On the conformation of proteins: Hydrophobic ordering of strands in β-pleated sheets. *J. Mol. Biol.* **1977**, *115*, 1–17. [CrossRef]
43. Sumner, A.K.; Nielsen, M.A.; Youngs, C.G. Production and Evaluation of Pea Protein Isolate. *J. Food Sci.* **1981**, *46*, 364–366. [CrossRef]
44. Shi, Y.; Mandal, R.; Singh, A.; Pratap Singh, A. Legume lipoxygenase: Strategies for application in food industry. *Legume Sci.* **2020**, *2*, e44. [CrossRef]
45. The UniProt Consortium. UniProt: The universal protein knowledgebase. *Nucleic Acids Res.* **2016**, *45*, D158–D169. [CrossRef]
46. Gatehouse, J.A.; Lycett, G.W.; Croy, R.R.; Boulter, D. The post-translational proteolysis of the subunits of vicilin from pea (*Pisum sativum* L.). *Biochem. J.* **1982**, *207*, 629–632. [CrossRef] [PubMed]
47. McClain, S.; Stevenson, S.E.; Brownie, C.; Herouet-Guicheney, C.; Herman, R.A.; Ladics, G.S.; Privalle, L.; Ward, J.M.; Doerrer, N.; Thelen, J.J. Variation in Seed Allergen Content From Three Varieties of Soybean Cultivated in Nine Different Locations in Iowa, Illinois, and Indiana. *Front. Plant Sci.* **2018**, *9*, 1025. [CrossRef] [PubMed]
48. Lam, A.C.Y.; Karaca, A.C.; Tyler, R.T.; Nickerson, M.T. Pea protein isolates: Structure, extraction, and functionality. *Food Rev. Int.* **2016**, *34*, 126–147. [CrossRef]
49. Zhao, X.; Sun, L.; Zhang, X.; Liu, H.; Zhu, Y. Effects of ultrafine grinding time on the functional and flavor properties of soybean protein isolate. *Colloids Surf. B Biointerfaces* **2020**, *196*, 111345. [CrossRef]
50. Kerr, W.L.; Ward, C.D.W.; McWatters, K.H.; Resurreccion, A.V.A. Effect of Milling and Particle Size on Functionality and Physicochemical Properties of Cowpea Flour. *Cereal Chem.* **2000**, *77*, 213–219. [CrossRef]
51. Vicente, J.; Pinto, J.; Menezes, J.; Gaspar, F. Fundamental analysis of particle formation in spray drying. *Powder Technol.* **2013**, *247*, 1–7. [CrossRef]
52. Taherian, A.R.; Mondor, M.; Labranche, J.; Drolet, H.; Ippersiel, D.; Lamarche, F. Comparative study of functional properties of commercial and membrane processed yellow pea protein isolates. *Food Res. Int.* **2011**, *44*, 2505–2514. [CrossRef]
53. Adebiyi, A.P.; Aluko, R.E. Functional properties of protein fractions obtained from commercial yellow field pea (*Pisum sativum* L.) seed protein isolate. *Food Chem.* **2011**, *128*, 902–908. [CrossRef]
54. Chao, D.F.; Aluko, R.E. Modification of the structural, emulsifying, and foaming properties of an isolated pea protein by thermal pretreatment. *CyTA J. Food* **2018**, *16*, 357–366. [CrossRef]
55. Gharsallaoui, A.; Cases, E.; Chambin, O.; Saurel, R. Interfacial and Emulsifying Characteristics of Acid-treated Pea Protein. *Food Biophys.* **2009**, *4*, 273–280. [CrossRef]
56. Dickinson, E. Protein Adsorption at Liquid Interfaces and the Relationship to Foam Stability. In *Foams: Physics, Chemistry and Structure*; Springer: London, UK, 1989; pp. 39–53. [CrossRef]
57. Denkov, N.; Marinova, K.; Denkov, N.; Marinova, K. Chapter 10—Antifoam effects of solid particles, oil drops and oil-solid compounds in aqueous foams. In *Colloidal Particles at Liquid Interfaces*; Binks, B.P., Horozov, T.S., Eds.; Cambridge University Press: Cambridge, UK, 2006; pp. 383–444. [CrossRef]
58. Murray, K.E.; Whitfield, F.B. The occurrence of 3-alkyl-2-methoxypyrazines in raw vegetables. *J. Sci. Food Agric.* **1975**, *26*, 973–986. [CrossRef]
59. Murat, C.; Bard, M.-H.; Dhalleine, C.; Cayot, N. Characterisation of odour active compounds along extraction process from pea flour to pea protein extract. *Food Res. Int.* **2013**, *53*, 31–41. [CrossRef]
60. Heng, L. *Flavour Aspects of Pea and Its Protein Preparations in Relation to Novel Protein Foods*; Wageningen University: Wageningen, The Netherlands, 2005.
61. García Arteaga, V.; Leffler, S.; Muranyi, I.; Eisner, P.; Schweiggert-Weisz, U. Sensory profile, functional properties and molecular weight distribution of fermented pea protein isolate. *Curr. Res. Food Sci.* **2021**, *4*, 1–10. [CrossRef] [PubMed]
62. Reynolds, A.G. 11—Viticultural and vineyard management practices and their effects on grape and wine quality. In *Managing Wine Quality*; Reynolds, A.G., Ed.; Woodhead Publishing: Cambridge, UK, 2010; pp. 365–444. [CrossRef]
63. Schlegel, K.; Leidigkeit, A.; Eisner, P.; Schweiggert-Weisz, U. Technofunctional and Sensory Properties of Fermented Lupin Protein Isolates. *Foods* **2019**, *8*, 678. [CrossRef]

Article

Assessment of the Physicochemical and Conformational Changes of Ultrasound-Driven Proteins Extracted from Soybean Okara Byproduct

Gilda Aiello [1], Raffaele Pugliese [2], Lukas Rueller [3], Carlotta Bollati [4], Martina Bartolomei [4], Yuchen Li [4], Josef Robert [3], Anna Arnoldi [4] and Carmen Lammi [4,*]

1 Department of Human Science and Quality of Life Promotion, Telematic University San Raffaele, 00166 Rome, Italy; gilda.aiello@unimi.it
2 NeMO Lab., ASST Grande Ospedale Metropolitano Niguarda, 20133 Milan, Italy; raffaele.pugliese@nemolab.it
3 Fraunhofer Institute for Environmental, Safety and Energy Technology UMSICHT, 46047 Oberhausen, Germany; Lukas.Rueller@umsicht.fraunhofer.de (L.R.); josef.robert@umsicht.fraunhofer.de (J.R.)
4 Department of Pharmaceutical Sciences, University of Milan, 20133 Milan, Italy; carlotta.bollati@unimi.it (C.B.); martina.bartolomei@unimi.it (M.B.); yuchen.li@unimi.it (Y.L.); anna.arnoldi@unimi.it (A.A.)
* Correspondence: carmen.lammi@unimi.it; Tel.: +39-025-0319-372

Citation: Aiello, G.; Pugliese, R.; Rueller, L.; Bollati, C.; Bartolomei, M.; Li, Y.; Robert, J.; Arnoldi, A.; Lammi, C. Assessment of the Physicochemical and Conformational Changes of Ultrasound-Driven Proteins Extracted from Soybean Okara Byproduct. *Foods* **2021**, *10*, 562. https://doi.org/10.3390/foods10030562

Academic Editors: Ute Schweiggert-Weisz and Emanuele Zannini

Received: 15 February 2021
Accepted: 3 March 2021
Published: 8 March 2021

Publisher's Note: MDPI stays neutral with regard to jurisdictional claims in published maps and institutional affiliations.

Abstract: This study was aimed at the valorization of the okara byproduct deriving form soy food manufacturing, by using ultrasound at different temperatures for extracting the residual proteins. The physicochemical and conformational changes of the extracted proteins were investigated in order to optimize the procedure. Increasing the temperature from 20 up to 80 °C greatly enhanced the yields and the protein solubility without affecting the viscosity. The protein secondary and tertiary structures were also gradually modified in a significant way. After the ultrasonication at the highest temperature, a significant morphological transition from well-defined single round structures to highly aggregated ones was observed, which was confirmed by measuring the water contact angles and wettability. After the ultrasound process, the improvement of peptides generation and the different amino acid exposition within the protein led to an increase of the antioxidant properties. The integrated strategy applied in this study allows to foster the okara protein obtained after ultrasound extraction as valuable materials for new applications.

Keywords: atomic force microscope; circular dichroism; phytic acid; green extraction; soybean proteins; soybean okara

1. Introduction

Soybean (*Glycine max*) is a protein-rich oilseed widely employed in the food industry for producing soy foods and beverages. Thanks to its nutrient content, soybean is used in several dishes as a valid alternative to meat, and it is added in various vegan-friendly food and beverages [1]. Soybean stands out not only for its nutritional value but also for the health benefits it provides (i.e., lowering of blood cholesterol level, increasing of bone density, and minimization of the risk of cancer development) [2].

The rising demand for plant-based foods is strengthening the growth of the soy food market across the globe. In fact, the global soy food market was worth 38.7 billion US$ in 2018 and is expected to reach the value of 53.1 billion US$ by 2024, registering a compound annual growth rate (CAGR) of around 5% during 2019–2024. In general, soybeans with a high protein content are chosen for the preparation of soymilk, compared to those utilized for oil extraction [3].

During soymilk and tofu production, soybeans are milled under hot (>80 °C) and alkaline (pH 8.0) conditions to guarantee a protein solubilization as well as to inactivate trypsin inhibitors and the enzyme lipoxygenase [4]. Insoluble materials are removed from the slurry using centrifugation: this process results in the production of a soy base, which is the precursor of soymilk or tofu, and a solid by-product generally named with the Japanese word okara, which contains about 50% dietary fibers, 25% protein, 10% lipid, and other nutrients [5,6] and is generally discarded or used as feed ingredient. Due to this interesting composition, okara byproducts have been already assessed for the extraction of fibers [7] and polysaccharides [8] as well as for the manufacture of snack [9]. Although it would be certainly advisable to extract the residual proteins for human use, this is impaired by the okara structure. A specific sustainable protein extraction methodology is thus required to reach this goal.

The ultrasound technology has been widely studied in the food industry for aiding the extraction of components of interest from plant starting materials [10–12]. The ultrasound technology shows promise as a green extraction technology, reasons including reductions in extraction times, less and more sustainable solvent use, and more effective energy utilization, as well as improvement of the quality of the product [13,14]. The success of ultrasound is attributed to the cavitation phenomenon. In fact, upon asymmetric bubble collapse, liquid jets are formed that can disrupt cells upon contact with cell walls, [15] causing the release of intracellular compounds.

Within the storage cells of soybean, protein is organized in 5–20 μm protein bodies, surrounded by a cytoplasmic network containing oil bodies in the size range of 0.2–0.5 μm stabilized by proteinaceous oleosins [16].

Different methods are applied for the protein extraction of soybean and okara. More specifically, the extraction of proteins is carried out using acid and alkaline conditions [17,18], enzyme assisted extractions [19,20], and more recently with the aid of ultrasonication technology [13], which is nevertheless still rarely employed. The complexity of soybean cellular microstructure influences the protein extraction in normal conditions, but thanks to the cavitation phenomenon, an improvement of extraction yields from hexane-defatted soy flakes is achievable at lab-scale. Protein functionality improvement from soybean protein isolate and concentrates has also been reported with positive results in protein solubility and particle size reduction [21]. In addition, the ultrasound treatment significantly increases the solubility of isolated soy proteins in water [22].

In light with these observations, this study is aimed at assessing the physicochemical and conformational changes of ultrasound-driven extracted proteins from soybean okara processing materials. To achieve this goal, the protein extraction was carried out using ultrasound at 20, 60, and 80 °C obtaining the samples named SoK_U20, SoK_U60, and SoK_U80, respectively. The high temperature limit was chosen taking into account that temperatures up to 80 °C are used in the processes where okara is produced as byproduct. The assessment of the changes in the protein profile, concentration, secondary and tertiary structure, hydrophobicity, solubility, antioxidant capacity, and rheological and morphological features was performed in comparison with untreated soy okara protein (SoK_nU). Herein, a combination of different techniques and an integrated strategy were applied in order to foster the okara protein samples obtained after ultrasound extraction as valuable and high-quality products for new applications and to provide, therefore, a more sustainable way to solve the environmental criticism related to the huge quantities of okara produced annually, which pose a significant disposal problem.

2. Materials and Methods

2.1. Chemicals and Reagents

All chemicals and reagents were of analytical grade and commercially available. More details are reported in Supplementary Information.

2.2. Samples and Ultrasonic System

The experimental investigations were performed with an ultrasonic batch system (TC 10, BSONIC GmbH, Germany) that allows power inputs of up to 4 kW at a frequency of 18 kHz and an oscillation amplitude of 45–60 µm. The probe tip has a diameter of 41.75 mm. Regulation of power input and recording of process parameters is done via PC. Berief Food GmbH, Germany, supplies the soy okara samples as frozen 3 kg units directly from the production process. For the liquid mixture of soy okara, tap water is used.

2.3. Ultrasound-Assisted Processing of Okara

Experiments of ultrasound-assisted extraction of proteins were performed at Fraunhofer UMSICHT, Germany, using the materials and ultrasonic system mentioned above. A defined ratio of 1:2.5 of soy okara and water was heated up to the experimental temperature of 20 °C (SoK_U20), 60 °C (SoK_U60), and 80 °C (SoK_U80), respectively, with continuous stirring. The ratio of 1:2.5 has been defined in previous experimental investigations, which are part of a separate publication. This ratio was selected as optimum concerning an effective protein extraction, the disintegration of okara, and the minimization of total treatment volume. The pH of the mixture was 6.5 (±0.14) at 20 °C. The ultrasound treatment was controlled and measured via PC and was conducted by immerging the ultrasound sonotrode into the beaker containing the samples (5 L). Ultrasound parameters of power input (4 kW), correlating oscillation amplitude (µm), and the treatment time of 0.5 min were fixed. Hence, the total energy input for all samples was 24 kJ/ L. These ultrasound treatment parameters have been defined ahead as part of a separate publication showing that they allow the best protein extraction yield. Directly after the ultrasound treatment, the processed samples were centrifuged at 12,298 *g* for 15 min (Avanti JXN-26, Beckmann Coulter) to separate solid and liquid fractions. Both fractions were separately freeze-dried (Alpha 2–4 LSC plus, Christ) in order to allow a simpler storage and delivery of the samples. The protein extraction yields were determined by Kjeldahl analysis.

2.4. Soy Okara Protein Extraction

Proteins from sonicated okara were extracted by modifying a method previously described [23]. More details are reported in Supplementary Materials.

2.5. Molecular Weight Distribution and MS Analysis

The molecular weight distribution of untreated and sonicated okara proteins was determined using reducing dodecyl sulfate-polyacrylamide gel electrophoresis (SDS-PAGE). The protein samples were prepared by mixing 15 µL of each sample with 10 µL of Laemmli buffer (4% SDS, 20% glycerol, 10%, 0.004% bromophenol blue, and 0.125 M Tris–HCl, pH 6.8). The mixtures were boiled for 5 min at 95 °C, and 25 µL of the mixture was loaded into each lane. The gel was composed of a 4% polyacrylamide stacking gel over a 12% resolving polyacrylamide gel. The electrophoresis was conducted at 100 V until the dye front reached the gel bottom. Staining was performed with colloidal Coomassie Blue and destaining with 7% (*v*/*v*) acetic acid in water. The gel image was acquired by using the Bio-Rad GS800 densitometer and analyzed by using the software quantity One 1-D. Gel bands for the SoK_U20, SoK_U60, and SoK_U80 lane were sliced, digested with trypsin according to a preceding paper, [23] and analyzed by nano-HPLC-CHIP-ESI Ion Trap using the same experimental conditions previously reported [24]. The MS data were analyzed by Spectrum Mill Proteomics Workbench (Rev B.04.00, Agilent), consulting the *Glycine max* (251326 entries) protein sequences database downloaded from the National Center for Biotechnology Information (NCBI) [25]. For MS/MS search the oxidation of methionine residues was set as variable modifications.

2.6. Peptidomic Profiles

The peptidomic profiles of the samples were obtained after ultrafiltration through 3 kDa molecular weight cut-off membranes (Amicon® Ultra, Millipore, Billerica, MA, USA). The recovered peptides were analyzed by nano LC–MS/MS analysis according to the chromatographic and MS condition reported in the material and methods. Figure S1 shows the MSn TIC of Sok_U80, Sok_U60, and Sok_U20. The MS data were analyzed by Spectrum

Mill Proteomics Workbench (Rev B.04.00, Agilent), consulting the *Glycine max* (251326 entries) protein sequences database downloaded from the National Center for Biotechnology Information (NCBI). For MS/MS analysis and searching against a polypeptide sequence database, a non-enzyme specific search considering all of the possible proteolytic cleavages was selected as criteria.

2.7. Protein Solubility and Water Binding Capacity

Protein solubility water binding capacity (WBC) were assessed according to a literature method with slight modifications [26]. More details are reported in Supplementary Materials.

2.8. Free Sulfhydryl Group Determination

Each sonicated sample (1 g) was dispersed into 9 mL deionized water to obtain the protein solution (1 wt%). Protein solutions were stirred for 2 h and centrifuged at 11,200 *g* for 20 min at 4 °C. The protein concentration in the samples before centrifugation and in the supernatants after centrifugation was determined according to the Bradford assay using BSA as a standard. The protein solubility was expressed as grams of soluble protein per 100 g of protein. All determinations were conducted in triplicate.

The water binding capacity (WBC) was assessed according to a literature method with slight modifications [26]. Briefly, 1 g of sample was dispersed in 10 mL distilled water in a 15 mL pre-weighed centrifuge tube. The dispersions were stirred for 30 min and then centrifuged at 7000 *g* for 25 min at room temperature. The supernatant was discarded, and the tubes were weighed to determine the amount of retained water per gram of sample.

2.9. Intrinsic Fluorescence Spectroscopy

The intrinsic fluorescence spectrum of each sample was obtained using a fluorescence spectrophotometer (Synergy H1, Biotek, Bad Friedrichshall, Germany). The samples were diluted in phosphate-buffered saline (PBS, 10 mM, pH 7.0) in order to reach the equal concentration of 0.05 mg/mL and transferred in Greiner UV-Star® 96 well plates flat bottom clear cyclic olefin copolymer (COC) wells (cycloolefine). The excitation wavelength was set as 280 nm, while the excitation and emission slit widths were each set as 5 nm. The emission wavelength range was set up from 300 to 450 nm, and the scanning speed was 10 nm/s.

2.10. Circular Dichroism (CD) Spectroscopy

CD spectra were recorded in continuous scanning mode (190–300 nm) at 25 °C using Jasco J-810 (Jasco Corp., Tokyo, Japan) spectropolarimeter. Following protocols reported in Supplementary Materials. The estimation of the peptide secondary structure was achieved by using a literature method [27].

2.11. Water Contact Angle Measurements

Contact angle measurements were performed on a Krüss Easy Drop instrument using freshly distilled water passed through a MilliQ apparatus. Sample powders were deposited on glass slides according to a literature procedure [28]. A 8 µL drop was produced and placed on the surface. Videos with 25 fps resolution were recorded. For each sample 2 to 3 measurements were performed, determining the first measurable contact angle and the total time needed for the droplet complete absorption.

2.12. Rheological Test

The rheological properties of soy okara proteins were tested using a stress/rate-controlled Kinexus DSR Rheometer (Netzsch) mounted with a parallel plate geometry (acrylic diameter, 20 mm; gap, 34 µm). All measurements were performed at a controlled temperature of 25 °C. The viscosity of okara proteins was measured using a flow step program, at increasing shear rate (0.01–1000 s^{-1}), to evaluate their non-Newtonian behavior. Afterwards, to evaluate the storage (G′) and loss moduli (G″), frequency sweep experiments were recorded at 0.1–100 Hz (strain 0.1%, in LVR). Each experiment was performed in triplicate. Data were processed using Origin^{TM} 8 software (Northampton, MA, US).

2.13. Atomic Force Microscopy (AFM)

AFM measurements were captured in tapping mode by using a Tosca system (Anton Paar) using single-beam silicon cantilever probes (Bruker RFESP–75 0.01–0.025 Ohm-cm Antimony (n) doped Si, cantilever f0, resonance frequency 75 kHz, constant force 3 N m^{-1}). AFM images were taken by depositing 3 µL solutions (final concentration of 0.1 mg mL^{-1}) onto freshly cleaved mica. The samples were kept on the mica for 5 min; subsequently, they were rinsed with distilled water to remove loosely bound peptides and then dried under ambient conditions for 24 h. The AFM morphological parameters were obtained using the Matlab-based open-source software FiberApp (Schmelzbergstrasse, Zurich).

2.14. Scanning Electron Microscopy (SEM)

Scanning electron microscopy (SEM) was performed with a Vega-3 microscope from TESCAN GmbH, Germany, at 20 kV acceleration voltage under high vacuum. To guarantee a high resolution, the images were taken with a secondary-electron detector (SE). Freeze-dried samples were fixed on a special two-sided adhesive tape and coated by a 10 nm gold surface in order to prevent electrostatic charging.

2.15. Determination of the Scavenging Activity by the DPPH Assay

The DPPH assay to determine the antioxidant activity in vitro was performed by a standard method with slight modifications [29]. Detailed information is reported in Supplementary Materials.

2.16. Phytic Acid (PA) Determination

Lyophilized samples were used for phytic acid determination, following the modified colorimetric method [30]. Aqueous phytic acid standards in concentrations of 0–100 µg/mL were used for quantification. Aliquots of 100 µL of samples and standards were diluted 25 times with 2.4 mL of H_2O; 600 µL of the diluted samples and standards were combined with 200 µL of modified Wade reagent (0.03% of $FeCl_36H_2O$ and 0.3% of sulfosalicylic acid), and the absorbance was measured at 500 nm.

2.17. Statistical Analysis

Data are presented as mean ± s.d. using GraphPad Prism 8 (GraphPad, La Jolla, CA, USA). Statistical analyses were carried out by t student test and ANOVA. p-values < 0.05 were considered significant.

3. Results and Discussions

3.1. Effect of Ultrasound Treatments on the Morphology of Soy Cells in Okara

Proteins are largely responsible for the main features of most foods, since their composition influences nutritional, rheological, and sensory properties. The new process may induce chemical modifications in the protein, impacting on the nutritional and technological features of the final products. The experimental setup used for the ultrasound-assisted processing of okara is illustrated in (Figure S1), while the parameters for the production of the samples are reported in (Table 1). Low ultrasound frequency at 18 kHz was used in order to guarantee an effective acoustic cavitation. In case of low frequencies, the periods of positive and negative pressure changes within the mechanical ultrasound wave are longer. Therefore, the growth process of the cavitation bubble is more effective, and higher bubble diameters can occur. Thus, the bubble implosion takes place with higher forces and results in a more disruptive cell disintegration in comparison to higher frequencies [13]. Accordingly, the oscillation amplitude is responsible for the net pressure change within the sonicated liquid. With higher amplitude, the pressure difference is higher, and acoustic cavitation bubbles are likely to increase in size. Therefore, a frequency of 18–22 kHz at a maximum oscillation amplitude is favorable for ultrasound disintegration and defines the ultrasound power input. Energy input is the product of power input and treatment time in relation to the sonicated volume. High energy input correlates with a high rate of disintegrated cells. However, treatment time should be minimized in order be able to convert the disintegration process to an industrial scale. Previous investigations that are part of a separate publication have shown that the selected ultrasound conditions guarantee an optimum of protein extraction and feasibility for potential industrial scale-up.

Table 1. Samples description.

Sample ID	T (°C)	Ultrasound	Treatment Time (min)	Energy Input (kJ/L)
SoK_nU	20	-	-	-
SoK_U20	20	4.0 kW	0.5	24
SoK_U60	60	4.0 kW	0.5	24
SoK_U80	80	4.0 kW	0.5	24

In order to evaluate the effect of ultrasonication coupled to the temperature gradient on the structure and morphology of soybean cells, scanning electron microscope analysis was carried out. Figure 1 shows that there are clear differences among the structures of the untreated and ultrasound treated samples. While both disrupted and intact cells co-exist in the sample treated with ultrasound at 20 °C (SoK_U20), the treatment at 60 °C (SoK_U60) and 80 °C (SoK_U80) produces visible disruptions of the cell structures. Since palisade-like cell structures contain a significant amount of protein bodies, an effective disruption of cells is indispensable for an efficient protein recovery (the protein extraction yields from SoK_U80, SoK_U60, and SoK_U20 were 23.5%, 14.9%, and 10.2%, respectively).

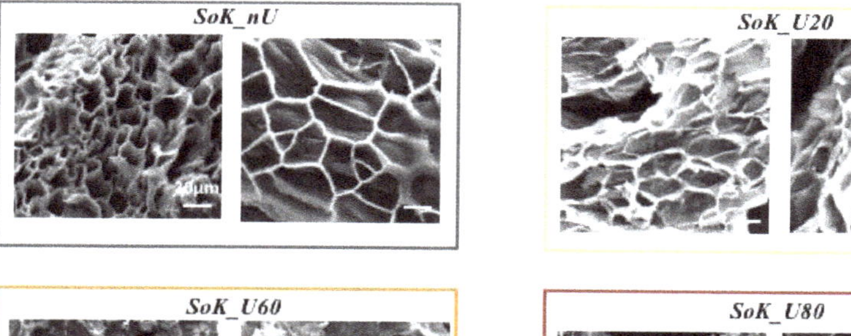

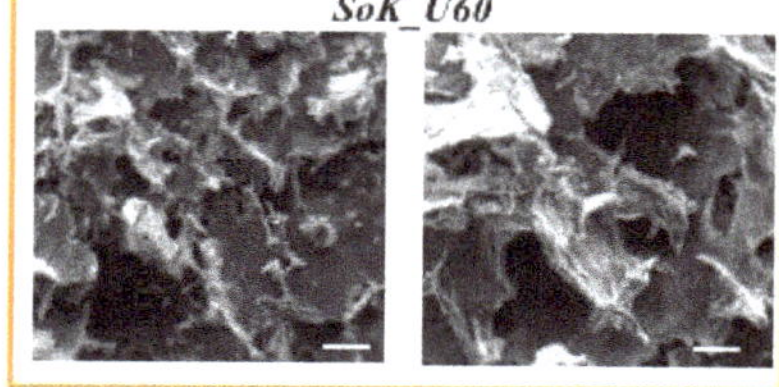

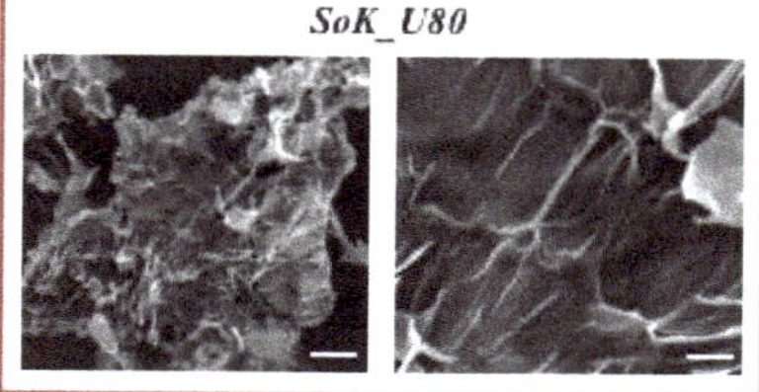

Figure 1. SEM of protein okara obtained with and without the ultrasound treatments. *Scheme 1600*x, r.: 4000x), Sok_U20 (ultrasonicated at 20 °C, magn.: l.: 800x, r.: 1600x), Sok_U60 (ultrasonicated at 60 °C, magn.: l.: 160x, r.: 400x) and Sok_U80 (ultrasonicated at 80 °C, magn.: l.: 400x, r.: 1600x).

3.2. Effect of Ultrasound Treatments on the Molecular Weight Distribution of Extracted Proteins

The effects of ultrasound treatments on the molecular properties of extracted proteins were explored by evaluating their molecular weight profile using reducing electrophoresis.

Figure 2A shows the SDS-PAGE profile of the untreated and sonicated protein samples. Under reducing conditions, four intense bands were observed for all the samples, with molecular weight ranges of 70–100 kDa, 40–55 kDa, 25–30 kDa, and ~18 kDa, respectively. The identified proteins for each band are reported in Table S1. Specifically, alpha and beta-subunits of conglycinin were identified at 70 kDa and 50 kDa, respectively. These findings are in line with those reported by other authors according to which no changes in the molecular weight profiles of squid mantle proteins, [31] walnut protein isolate [32], and soybean proteins were observed after sonication [33].

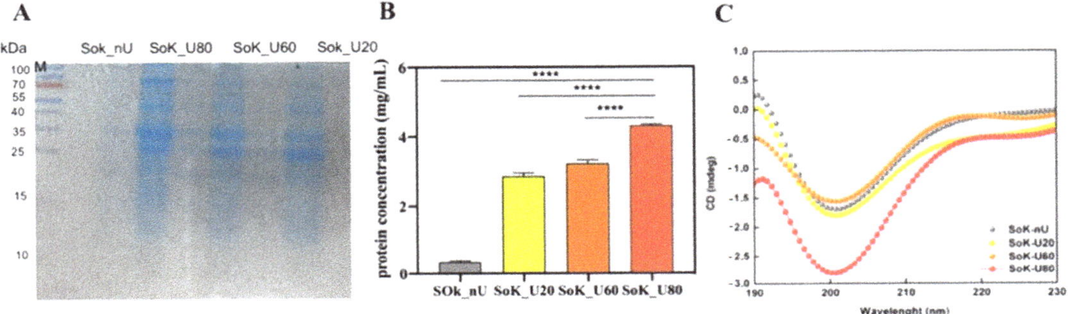

Figure 2. Protein profile, concentration, and CD spectra. (**A**) Reduced SDS-PAGE protein profile of ultrasonicated and untreated proteins (M, pre-stained molecular marker). (**B**) Determination of the protein concentration by the Bradford assay. (**C**) CD spectra of okara proteins. SoK_nU (untreated), Sok_U20 (ultrasonicated at 20 °C), Sok_U60 (ultrasonicated at 60 °C), and Sok_U80 (ultrasonicated at 80 °C).

The intensity of the electrophoretic bands of sonicated samples at different temperature was greater than that of the untreated sample, which may be attributed to the greater water-solubility of the sonicated proteins. This evidence was confirmed also by Bradford assay according to which the detected amounts of protein were 4.31 ± 0.03, 3.22 ± 0.01, 2.85 ± 0.1, and 0.36 ± 0.04 mg/mL for SoK_80, SoK_60, SoK_20, and SoK_nU, respectively. Therefore, the ultrasound process performed at room temperature (SoK_U20) led to an improvement of protein extraction yield by up to 6.9-fold versus SoK_nU (Figure 2B). This is in agreement with other studies that have shown that ultrasound improves protein extraction yield from soybeans in a lab-scale system [34].

The study on the impact of the temperature on the ultrasound aided protein extraction indicated that in SoK_U60 and SoK_U80, the protein extraction yields were increased by 13.0% and 51.2%, respectively, vs SoK_U20. This high recovery yield was in agreement with SEM investigation that highlighted the progressive destruction of the cell structure in ultrasound treated samples as a function of the temperature leading to an improvement of released proteins.

After 3 kDa cut-off filtration, each sample was submitted to a peptidomic investigation. The MS/MS analysis revealed that the ultrasonication induced a progressively greater peptide release that proportional to the increasing temperature (Figure S2). In agreement with the degree of hydrolysis (data not shown), the peptide sequences identified after ultrasound treatments increased also as a function of the temperature (Table S2): in fact, 24, 30, and 37 different peptides were identified in Sok_U20, Sok_U60, and Sok_U80 samples, respectively. Interestingly, in all samples the peptide lengths were similar, ranging from 6 to 29 amino acid residues, and the pI values were comprised between 3.8 and 9.9.

3.3. Circular Dichroism (CD) of Okara Proteins

To investigate the effect of ultrasound treatment and temperature combination on the secondary structure of extracted proteins, CD spectra in the far UV region of 190–230 were recorded (Figure 2C). One positive and one negative Cotton effect at 193 and 200 nm, respectively, were observed for SoK_nU, suggesting an α-helix rich conformation. Increasing the temperature of the ultrasound treatment from 20 and 60 °C, the intensity of the Cotton effect peaks was greatly decreased, suggesting a structural transition from α-helix into β-sheet rich conformations. The latter conformation was clearly visible in the SoK_U80 sample, where one maximum peak at 195 nm and one minimum peak at 200 nm were observed, demonstrating a redshift of the maximum peak, thus indicating a highly ordered β-sheet rich structures after heating the sample at 80 °C (Figure 2C).

To gain further information about the secondary structure of soybean okara proteins, the Raussens and coworkers' tool was applied [27]. Results, summarized in (Table 2), suggest that a reduction of the percentage of α-helices and an improvement of β-sheet were obtained for the proteins after ultrasound treatment (Sok_U20) versus the untreated sample (SoK_nU). Clearly the increase of temperature coupled to ultrasonication induced a significant secondary structure variation. Overall, reductions of α-helices up to 11.3% and 1.1% at 60 and 80 °C, respectively, vs SoK_U20 (26.7%) were observed. In addition, increases of β-sheet up to 33.2% and 37.7% at 60 and 80 °C, respectively, vs SoK_U20 (28.1%) were observed. An overall slight improvement of random coil was also observed for each experimental condition.

Table 2. Percentage of secondary structure composition of soybean okara proteins.

Secondary Structure	SoK_nU	SoK_U20	SoK_U60	SoK_U80
Helix (%)	32.4	26.7	11.3	1.1
Beta (%)	18.5	28.1	33.2	37.7
Turn (%)	12.5	12.5	12.5	12.5
Random (%)	37.6	38.2	39.5	42.0

3.4. Free-Sulfhydryl Group (SH) Content

Sulfhydryl groups (SH) and disulfide bonds (S-S) are important chemical bonds that stabilize the conformation of protein molecules and play very important roles in functional properties, such as foaming and emulsifying abilities. The measurement of the content of free-SH groups located on the surface of okara proteins was used to provide further insights into the ability of sonication to cause changes in the protein tertiary structure.

Figure 3A shows significant reductions in free SH content of the extracted proteins after sonication. In detail, the free SH contents of SoK_nU was 59.4 ± 3.1 μmol/g and the ultrasonication at 20 °C led to a reduction of the free SH content up to 7.4 ± 0.6 μmol/g. These findings clearly indicate that the ultrasonication determines a significant effect on the structure of the protein with a reduction of the exposition of the hydrophobic amino acids containing thiol groups, probably due to the formation of intermolecular disulfide bonds (S-S), which modulates the folding of the extracted proteins. Moreover, the findings suggest that an increment of the temperature induced an additional decrease of the free SH content by 28.4% and 60.8% for SoK_U60 (5.3 ± 0.8 μmol/g) and SoK_U80 (2.9 ± 0.6 μmol/g), respectively, versus SoK_U20. These data underline the effects of two different factors: the ultrasonication process and the increasing temperature that induces extensive modifications of the protein structures. This phenomenon may be possibly due to the generation of radical species during the sonication process, which may oxidize susceptible functional groups such the thiol group, leading to the formation of intermolecular disulfide bonds (S-S). Indeed, the thermolysis induced by cavitation may produce hydroxyl radicals and hydrogen atoms that can induce the formation of radical species [35]. However, other researchers have reported that the sonication can increase the free SH content of egg and soy proteins [36]. The protein type, the solutions, and the processing conditions may produce these different outcomes.

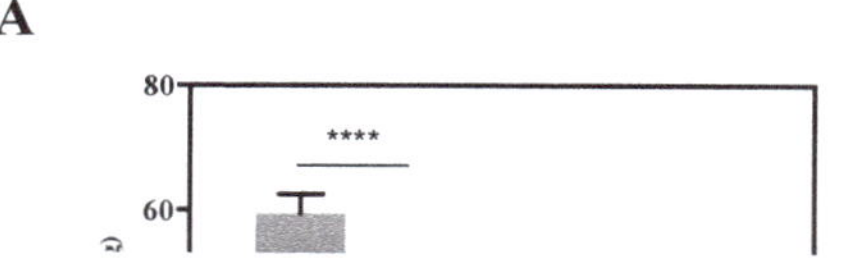

3.5. Intrinsic Fluorescence

More information regarding the effect of ultrasound at different temperatures on the protein structural changes was obtained by applying intrinsic fluorescence spectroscopy, a technique that can be used to monitor alterations in protein tertiary structure due to the sensitivity of the protein amino acid residues to the polarity of the microenvironment [37]. Since the intrinsic fluorescence is mainly due to the presence of tryptophan (Trp) and tyrosine (Tyr) residues, which have strong fluorescence quantum yield, after excitation at 280 nm, the fluorescence spectrum of each sample was recorded in the wavelength range 310–450 nm. An improvement of fluorescence intensity was detected as a function of the temperature reached during the ultrasound-assisted protein extraction from okara (Figure 3B). The SoK_U20 had a significant 5.7% increase of fluorescence intensities compared to the untreated sample (SoK_nU) whereas, the fluorescent intensity of SoK_U60 and SoK_U80 was increased by 43.5% and 103.2%, respectively, compared to SoK_U20. The increased fluorescence intensity is correlated to an increase in the number of exposed Trp residues. Similar trends have been observed also in soybean and chicken plasma proteins submitted to ultrasound extraction [21,36]. In particular, a recent paper has shown that in soybean all sonication conditions induce a significant 13–41% increase of the fluorescence intensities compared to the untreated sample [21].

The changes of protein tertiary structure may be determined by monitoring fluorescence intensity at the maximum wavelength (λ_{max}) [38]. In these experiments, SoK_U60 reached the λ_{max} at 332 nm, whereas all the other samples displayed a λ_{max} at 340 nm. Trp resides can be classified into three types based on their different λ_{max} values [39]: (i) buried Trp residue at λ_{max} between 330 and 332 nm, (ii) exposed Trp with limited water contact at λ_{max} between 340 and 342 nm, and (iii) exposed Trp residue at λ_{max} between 350 and 353 nm. Hence, Trp is exposed with limited water contact in SoK_nU, SoK_U20, and SoK_U80, whereas it is buried in SoK_U60. The slight shift in λ_{max} from 332 nm to 340 nm of SoK_U60 indicates a different behavior of this sample, since the Trp residues have shifted from being exposed with limited water contact to be buried.

3.6. Protein Hydrophobicity After Ultrasound Reatments

In order to assess the variation of the hydrophobicity, the water contact angles of the samples were measured (Figure 3C,D). Briefly, sample powders were deposited on glass slides and a drop of water was produced and fell on the surface. As shown in the videos (see Supplementary Materials), the proteins extracted with ultrasound coupled to the temperature gradient showed an improved ability to absorb the water drop in comparison with the untreated sample (SoK_nU). Moreover, SoK_nU has a slower ability to absorb water and an improved ability to swell. On the contrary, SoK_U20, SoK_U60, and SoK_U80 absorbed the water drop faster without swelling (Figure 3C,D), suggesting an improvement of their wettability. Precisely, the water contact angles (θ) of SoK_U20, SoK_U60, and SoK_U80 were the 42.7 ± 2.1%, 67.2 ± 4.5% and 68.9 ± 5.2%, respectively, versus SoK_nU (Figure 3C). Therefore, all values were smaller than that of the untreated sample, but slight improvements were detected increasing the temperature of the ultrasound treatment. Notably, the water drop was absorbed in 16 s by SoK_nU, whereas in 0.6 s, 0.9 s, and 1.2 s by SoK_U20, SoK_U60, and SoK_ U80, respectively (Figure 3D).

3.7. Morphological Analysis

To investigate the effect of the ultrasound and heat treatments on the extracted proteins, the morphologies of the samples were studied by AFM (Figure 4A). Well-defined single round structures were observed in SoK_nU alone, with an average height of 1.6 ± 0.9 nm (Figure S2) and no obvious changes of aggregates were found in SoK_U20 (2.0 ± 1.7 nm). Instead, SoK_U60 showed unevenly distributed aggregated structures, with a slightly increased height compared to SoK_nU and SoK_U20, namely, 5.6 ± 1.7 nm. Conversely, in sample SoK_U80, a significant morphological transition from well-defined single round structures to highly aggregated structures was observed. The size of the structures was increased compared to that of SoK_nU alone and SoK_U20, with a height of 47 ± 10.9 nm, indicating that the ultrasound coupled with an 80 °C temperature strongly contributed to the formation of these large aggregates.

These data are in good agreement with those obtained through the turbidity assay (Figure 4B), where an increase in turbidity values was observed following heating of the samples, suggesting that the higher temperature led to a significant increase in the size of the protein aggregates.

In addition, rheological experiments were performed to investigate the storage (G′) and loss (G″) moduli in the function of angular frequency (1–100 Hz). All samples exhibited a G′/G″ profile almost unchanged along the tested frequency range (Figure 4C). However, the ultrasound treated samples displayed higher G′ values (≈25 Pa) compared to the untreated sample (SoK_nU, G′ = 0.2 Pa); this behavior may indicate different networks of nano/microstructures inside the extracted protein, thus potentially influencing their mechanical properties.

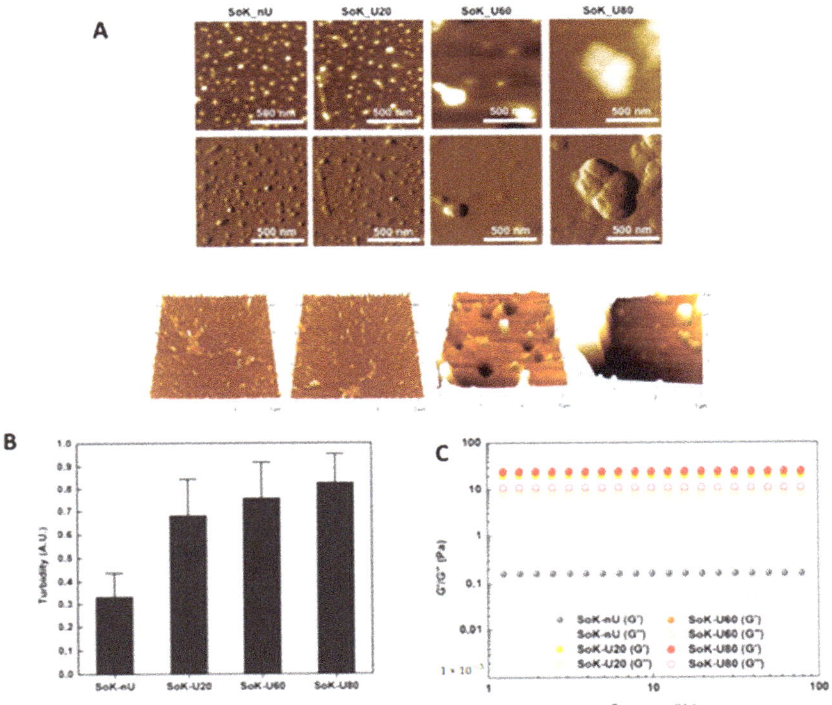

Figure 4. (**A**) Morphological organization of extracted proteins. (**B**) Turbidity of okara proteins measured at 405 nm. (**C**) Frequency-depended oscillatory rheology (1–100 Hz) of extracted proteins at fixed strain 0.1%. SoK_nU (untreated), Sok_U80 (ultrasonicated at 80 °C), Sok_U60 (ultrasonicated at 60 °C), and Sok_U20 (ultrasonicated at 20 °C).

3.8. Protein Solubility, Water Binding Capacity (WBC), and Viscosity of Extracted Proteins

The ultrasonication process affected also the protein solubility, which represents a good index of protein functionality [39]. This feature reflects protein denaturation and aggregation, which modulate many important functional properties, such as emulsification, solubility, gelation, and viscosity [40]. The solubility of untreated and treated samples by ultrasound is shown in Figure 5A. The protein solubility of SoK_U20 is 6.5-fold higher (38.2 ± 1.3 mg protein/g biomass SoK_U20) than the untreated sample (5.9 ± 0.1 mg protein/g biomass). In addition, the temperature during the ultrasound process proportionally increases the protein solubility that for SoK_U60 (59.6 ± 1.2 mg protein/g biomass) and SoK_U80 (86.1 ± 1.1 mg protein/g biomass) were greater by 1.5 and 2.3 folds, respectively, than that of SoK_U20 (Figure 5A). This evidence is supported by the fact the during ultrasonication the cavitation bubbles induce the unfolding of proteins causing an increased exposure of hydrophilic amino acid residues towards water thus contributing to the formation of soluble protein aggregates [41]. In addition, the increase of peptides within the samples may also contribute to the improvement of solubility.

An important property of proteins is the ability to interact with water, which influences their propensity to form gels, to dissolve, to swell, and to act as stabilizer in emulsions [42]. The measurement of their WBC is a conventional way to describe the interaction of proteins with water. Therefore, in order to evaluate the effect on WBC, dedicated experiments were carried out (Figure 5B). The WBC of the proteins obtained with the classical extraction method (SoK_nU) was 7.2 ± 0.01 g H_2O/g protein (Figure 5B), in line with a previous study, which compared the WBC of soybean, pea, and lupin isolated proteins, reporting that the specific WBC of soybean proteins is 9.2 g H_2O/g protein respectively [43]. Moreover, the ultrasonication at 20 °C (SoK_U20) did not produce a significant effect on the WBC (7.0 ± 0.01 g H_2O/g protein). On the contrary, a significant variation of WBC was observed after ultrasound extraction at 60 and 80 °C: the WBC's of SoK_U60 and SoK_U80 were 18 ± 0.5% and 31.4 ± 0.8% higher than SoK_U20, respectively (Figure 5B).

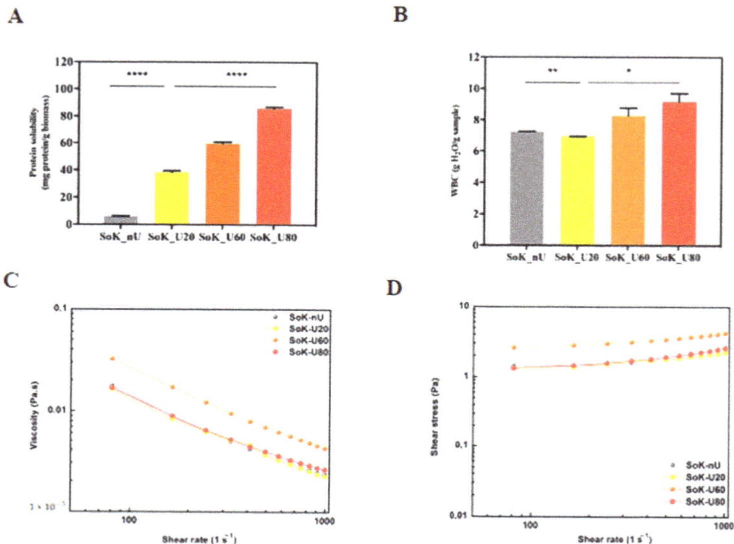

Figure 5. Solubility and rheological properties. (**A**) Protein solubility. (**B**) Water binding capacity (WBC). (**C**) Viscosity measurements at increasing shear rate. (**D**) Shear stress (σ) measurements at increasing shear rate of extracted proteins confirmed the non-Newtonian shear-thinning behavior of untreated and sonicated proteins. SoK_nU (untreated), Sok_U80 (ultrasonicated at 80 °C), Sok_U60 (ultrasonicated at 60 °C), and Sok_U20 (ultrasonicated at 20 °C). Statistical analysis was performed by one-way ANOVA; (*) $p < 0.5$, (**) $p < 0.01$, and (****) $p < 0.0001$. The data are represented as the means ± s.d. of three independent experiments.

In light of these observations, rheology was employed to evaluate the viscous properties: All samples displayed a non-Newtonian shear-thinning behavior with a decrease of viscosity that was concomitant with the shear-rate increase (Figure 5C). Even if the SoK_U60 showed an increased viscosity (0.032 Pa.s) in respect to samples SoK_nU (0.017 Pa.s), SoK_U20 (0.016 Pa.s), and SoK_U80 (0.016 Pa.s), all samples had negligible differences at higher shear rate values (700–1000 s^{-1}). The non-Newtonian shear-thinning behavior of all samples was also confirmed by assessing the shear stress (σ) trend alongside shear-rate increments (Figure 5D).

3.9. *In Vitro Antioxidant Activity Assayed by DPPH*

The DPPH radical scavenging assay is one of the most commonly used single electron transfer (SET) based antioxidant procedures. Each sample was tested at the concentration of 0.1 mg/Ml. A comparison of the result of the untreated sample with that of the sample treated with ultrasound at room temperature shows that this treatment increases the ability of the sample to scavenge the DPPH radical: in fact, Sok_U20 diminished the DPPH radicals by 90 ± 5.6% compared to SoK_nU ($p < 0.0001$) (Figure 6A). In addition, further improvements of the antioxidant capacity are induced by the thermal treatments, since SoK_U60 and SoK_U80 reduced the DPPH radical by 56.3 ± 0.6% and 72.2 ± 0.3%, respectively ($p < 0.0001$, Figure 6B). The increased antioxidant capacity may be due to the exposure of hidden amino acid residues and side chains with antioxidant capacities (which are usually hidden within the three-dimensional structure of protein molecules). However, it is also important to underline that the peptidomic analysis had already shown that a higher temperature during the ultrasound treatment has the consequence of a relevant increment of the presence of short peptides that may have additional roles in the detected antioxidant activity. Moreover, previous studies carried out on different food matrices (soybean included) have demonstrated that an improved antioxidant ability might be attributed to the formation of short-chain peptides induced by the ultrasound treatment [21].

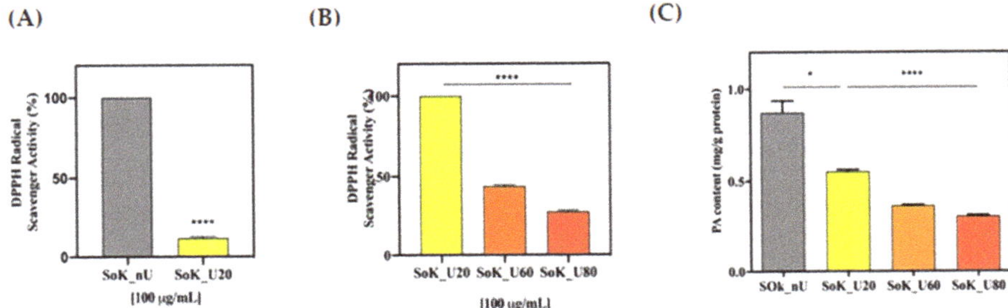

Figure 6. (**A**,**B**) Antioxidant evaluation of SoK_nU, SoK_U20, SoK_U60, and SoK_U80 by DPPH assay; (**C**) PA content determination. SoK_nU (untreated), Sok_U80 (ultrasonicated at 80 °C), Sok_U60 (ultrasonicated at 60 °C), and Sok_U20 (ultrasonicated at 20 °C). Statistical analysis was performed by one-way ANOVA (*) $p < 0.5$, (****) $p < 0.0001$. The data are represented as the means ± s.d. of three independent experiments.

3.10. Phytic Acid Reduction by Ultrasound Treatment

Since literature indicates that ultrasounds may be successfully applied to reduce the anti-nutritional factor phytic acid [44], it was decided to investigate also this aspect. By normalizing the content of phytic acid in respect to the protein content of each sample, our finding showed that the ultrasonication coupled to temperatures simultaneously reduced the phytic acid proteins interactions compared to the raw sample as reported in (Figure 6C). In detail, a reduction of phytic acid content by 37.0 ± 1.0% was observed in SoK_U20 (0.5 ± 0.01 mg/g of protein) versus SoK_nU (0.9 ± 0.06 mg/g of protein) ($p < 0.05$, Figure 6C). This may be explained considering that the acoustic effect of cavitation leads to a disruption of phytic acid, mainly localized in bran layer of soybean okara, by increasing the area and extraction rate of phytic acid into the solvent. Furthermore, the increase of temperature leads to an additional significant reduction of the phytic acid content compared to the ultrasound extraction performed at 20 °C. In fact, in SoK_U60 (0.36 ± 0.03 mg/g of protein) and SoK_U80 (0.30 ± 0.05 mg/g of protein), the phytic acid content decreased by 28 ± 3% and 40 ± 5%, respectively, versus Sok_U20 (0.5 ± 0.01 mg/g of protein) ($p < 0.0001$, Figure 6C).

4. Conclusions

The effect of ultrasound on the separation and extraction has been extensively studied: It is widely accepted that this process intensifies the extraction of valuable components from soybeans, leading to an overall improvement of protein yield. In this context, our investigation confirms that the ultrasound assisted extraction coupled to a gradient of temperature is also a useful strategy to improve the recovery of proteins from soy okara by-products. From an economical and environmental point of view, these findings contribute to fostering the soy okara protein extracted by ultrasonication processing as valuable and high-quality products for new applications, thus providing a sustainable way to solve the environmental criticism related to the huge quantities of okara produced annually.

An overall consideration of our results permits to conclude that the ultrasound procedure coupled to a temperature gradient modifies in a significant way the protein secondary and tertiary structures. In fact, a local reduction of α-helices structures and an improvement of beta-sheets and random coil conformations were observed depending on the applied temperature. In addition, a reduction of the free thiol groups and a different distribution of Trp were also detected within the protein samples. The AFM analysis demonstrated a significant morphological transition from well-defined single round structures to highly aggregated ones after the ultrasonication at the higher temperature, suggesting that these aggregates possessed more hydrophilic surfaces and more hydrophobic cores than the untreated sample. This feature was confirmed by measuring the water contact angle, whose results clearly indicated that untreated samples were more hydrophobic than the treated ones, a fact further confirmed by the slower ability of the untreated sample to absorb water drop than ultrasound extracted proteins. All these results were in agreement with the enhancement of the protein yields induced by the ultrasound treatments at different temperatures.

The improvement of protein yields may have a twofold explanation: (a) the cavitation phenomenon induced by ultrasound process enhances the disruption of intact cells leading to an increased extracellular release; (b) the improvement of protein solubility has the consequence of an increase of the recovery. Notably, in this ultrasound extraction, the improvements of protein solubility and water binding capacity appear to depend on the temperature in a similar way. On the contrary, rheological experiments do not support any variation of protein viscosity.

Finally, from a functional point of view, the improvement of peptides generation and the different amino acid exposition within the protein after the ultrasound process led to an increase of the antioxidant properties of the samples and to a reduction of their PA content.

Supplementary Materials: The following are available online at https://www.mdpi.com/2304-8158/10/3/562/s1. Table S1: Protein identified in SoK_nU, SoK_U80, SoK_U60, SoK_U20 after tryptic digestion. Table S2: Peptide sequences identified in Sok_U80, Sok_U60, Sok_U20, respectively after ultrasound treatment. Figure S1: Experimental setup of ultrasound-assisted extraction. Figure S2: Total ion chromatograms (TICs) of (A) Sok_U80; (B) Sok_U60 and (C) Sok_U20, respectively; Videos 1, 2,3, and 4: wettability of SoK_nU, SoK_U20, SoK_U60, and SoK_U80, respectively.

Author Contributions: Experiments, C.L., R.P., L.R., G.A., C.B., M.B., Y.L.; discussion of the results, C.L. manuscript writing (original draft), C.L., R.P., G.A., L.R.; manuscript editing C.L., J.R. and A.A. All authors have read and agreed to the published version of the manuscript.

Funding: This transnational project is part of the ERA-Net SUSFOOD2 with funding provided by Italian national/regional sources and co-funding by the European Union's Horizon 2020 research and innovation program. Part of this work was also financially supported by the German Federal Ministry of Food and Agriculture (BMEL) through the Federal Office for Agriculture and Food (BLE), grant number 2818ERA12B.

Data Availability Statement: The data presented in this study are available in this article and Supplementary Materials.

Conflicts of Interest: The authors declare no competing financial interest.

References

1. Rizzo, G.; Baroni, L. Soy, Soy Foods and Their Role in Vegetarian Diets. *Nutrients* **2018**, *10*, 43. [CrossRef]
2. Messina, M.; Messina, V. The Role of Soy in Vegetarian Diets. *Nutrients* **2010**, *2*, 855–888. [CrossRef]
3. Soy Food Market: Global Industry Trends, S., Size, Growth, Opportunity and Forecast 2019–2024. Available online: https://www.researchandmarkets.com/reports/4828757/soy-food-market-global-industry-trends-share (accessed on 1 January 2021).
4. Vishwanathan, K.H.; Singh, V.; Subramanian, R. Wet grinding characteristics of soybean for soymilk extraction. *J. Food Eng.* **2011**, *106*, 28–34. [CrossRef]

5. Li, B.; Qiao, M.Y.; Lu, F. Composition, Nutrition, and Utilization of Okara (Soybean Residue). *Food Rev. Intern.* **2012**, *28*, 231–252. [CrossRef]
6. Colletti, A.; Attrovio, A.; Boffa, L.; Mantegna, S.; Cravotto, G. Valorisation of By-Products from Soybean (Glycine max (L.) Merr.) Processing. *Molecules* **2020**, *25*, 2129. [CrossRef] [PubMed]
7. Fan, X.; Chang, H.; Lin, Y.; Zhao, X.; Zhang, A.; Li, S.; Feng, Z.; Chen, X. Effects of ultrasound-assisted enzyme hydrolysis on the microstructure and physicochemical properties of okara fibers. *Ultrason. Sonochem.* **2020**, *69*, 105247. [CrossRef]
8. Villanueva-Suárez, M.J.; Pérez-Cózar, M.L.; Redondo-Cuenca, A. Sequential extraction of polysaccharides from enzymatically hydrolyzed okara byproduct: Physicochemical properties and in vitro fermentability. *Food Chem.* **2013**, *141*, 1114–1119. [CrossRef] [PubMed]
9. Katayama, M.; Wilson, L.A. Utilization of okara, a byproduct from soymilk production, through the development of soy-based snack food. *J. Food Sci.* **2008**, *73*, S152–S157. [CrossRef] [PubMed]
10. Chemat, F.; Zill e, H.; Khan, M.K. Applications of ultrasound in food technology: Processing, preservation and extraction. *Ultrason. Sonochem.* **2011**, *18*, 813–835. [CrossRef]
11. Esclapez, M.D.; Garcia-Perez, J.V.; Mulet, A.; Carcel, J.A. Ultrasound-Assisted Extraction of Natural Products. *Food Eng. Rev.* **2011**, *3*, 108–120. [CrossRef]
12. Shirsath, S.R.; Sonawane, S.H.; Gogate, P.R. Intensification of extraction of natural products using ultrasonic irradiations A review of current status. *Chem. Eng. Process. Process Intensif.* **2012**, *53*, 10–23. [CrossRef]
13. Chemat, F.; Rombaut, N.; Sicaire, A.G.; Meullemiestre, A.; Fabiano-Tixier, A.S.; Abert-Vian, M. Ultrasound assisted extraction of food and natural products. Mechanisms, techniques, combinations, protocols and applications. A review. *Ultrason. Sonochem.* **2017**, *34*, 540–560. [CrossRef] [PubMed]
14. Sicaire, A.G.; Vian, M.A.; Fine, F.; Carre, P.; Tostain, S.; Chemat, F. Ultrasound induced green solvent extraction of oil from oleaginous seeds. *Ultrason. Sonochem.* **2016**, *31*, 319–329. [CrossRef] [PubMed]
15. Li, H.Z.; Pordesimo, L.; Weiss, J. High intensity ultrasound-assisted extraction of oil from soybeans. *Food Res. Int.* **2004**, *37*, 731–738. [CrossRef]
16. Rosenthal, A.; Pyle, D.L.; Niranjan, K. Simultaneous aqueous extraction of oil and protein from soybean: Mechanisms for process design. *Food Bioprod. Process.* **1998**, *76*, 224–230. [CrossRef]
17. Ma, C.Y.; Liu, W.S.; Kwok, K.C.; Kwok, F. Isolation and characterization of proteins from soymilk residue (okara). *Food Res. Int.* **1996**, *29*, 799–805. [CrossRef]
18. Vishwanathan, K.H.; Govindaraju, K.; Singh, V.; Subramanian, R. Production of okara and soy protein concentrates using membrane technology. *J. Food Sci.* **2011**, *76*, E158–E164. [CrossRef]
19. Sari, Y.W.; Bruins, M.E.; Sanders, J.P.M. Enzyme assisted protein extraction from rapeseed, soybean, and microalgae meals. *Ind. Crops Prod.* **2013**, *43*, 78–83. [CrossRef]
20. Fischer, M.; Kofod, L.V.; Schols, H.A.; Piersma, S.R.; Gruppen, H.; Voragen, A.G.J. Enzymatic extractability of soybean meal proteins and carbohydrates: Heat and humidity effects. *J. Agric. Food. Chem.* **2001**, *49*, 4463–4469. [CrossRef]
21. Tian, R.; Feng, J.; Huang, G.; Tian, B.; Zhang, Y.; Jiang, L.; Sui, X. Ultrasound driven conformational and physicochemical changes of soy protein hydrolysates. *Ultrason. Sonochem.* **2020**, *68*, 105202. [CrossRef]
22. Jambrak, A.R.; Lelas, V.; Mason, T.J.; Kresic, G.; Badanjak, M. Physical properties of ultrasound treated soy proteins. *J. Food Eng.* **2009**, *93*, 386–393. [CrossRef]
23. Aiello, G.; Fasoli, E.; Boschin, G.; Lammi, C.; Zanoni, C.; Citterio, A.; Arnoldi, A. Proteomic characterization of hempseed (*Cannabis sativa* L.). *J. Proteomics* **2016**, *147*, 187–196. [CrossRef]
24. Aiello, G.; Lammi, C.; Boschin, G.; Zanoni, C.; Arnoldi, A. Exploration of Potentially Bioactive Peptides Generated from the Enzymatic Hydrolysis of Hempseed Proteins. *J. Agric. Food. Chem.* **2017**, *65*, 10174–10184. [CrossRef] [PubMed]
25. Lammi, C.; Arnoldi, A.; Aiello, G. Soybean Peptides Exert Multifunctional Bioactivity Modulating 3-Hydroxy-3-Methylglutaryl-CoA Reductase and Dipeptidyl Peptidase-IV Targets in Vitro. *J. Agric. Food. Chem.* **2019**, *67*, 4824–4830. [CrossRef] [PubMed]
26. Malomo, S.A.; He, R.; Aluko, R.E. Structural and functional properties of hemp seed protein products. *J. Food Sci.* **2014**, *79*, C1512–C1521. [CrossRef] [PubMed]
27. Raussens, V.; Ruysschaert, J.M.; Goormaghtigh, E. Protein concentration is not an absolute prerequisite for the determination of secondary structure from circular dichroism spectra: A new scaling method. *Anal. Biochem.* **2003**, *319*, 114–121. [CrossRef]
28. Nowak, E.; Combes, G.; Stitt, E.H.; Pacek, A.W. A comparison of contact angle measurement techniques applied to highly porous catalyst supports. *Powder Technol.* **2013**, *233*, 52–64. [CrossRef]
29. Lammi, C.; Bellumori, M.; Cecchi, L.; Bartolomei, M.; Bollati, C.; Clodoveo, M.L.; Corbo, F.; Arnoldi, A.; Nadia, M. Extra Virgin Olive Oil Phenol Extracts Exert Hypocholesterolemic Effects through the Modulation of the LDLR Pathway: In Vitro and Cellular Mechanism of Action Elucidation. *Nutrients* **2020**, *12*, 1723. [CrossRef]
30. Gao, Y.; Shang, C.; Maroof, M.A.S.; Biyashev, R.M.; Grabau, E.A.; Kwanyuen, P.; Burton, J.W.; Buss, G.R. A modified colorimetric method for phytic acid analysis in soybean. *Crop Sci.* **2007**, *47*, 1797–1803. [CrossRef]
31. Higuera-Barraza, O.A.; Torres-Arreola, W.; Ezquerra-Brauer, J.M.; Cinco-Moroyoqui, F.J.; Figueroa, J.C.R.; Marquez-Rios, E. Effect of pulsed ultrasound on the physicochemical characteristics and emulsifying properties of squid (Dosidicus gigas) mantle proteins. *Ultrason. Sonochem.* **2017**, *38*, 829–834. [CrossRef]

32. Zhu, Z.B.; Zhu, W.D.; Yi, J.H.; Liu, N.; Cao, Y.G.; Lu, J.L.; Decker, E.A.; McClements, D.J. Effects of sonication on the physicochemical and functional properties of walnut protein isolate. *Food Res. Int.* **2018**, *106*, 853–861. [CrossRef]
33. Hu, H.; Li-Chan, E.C.Y.; Wan, L.; Tian, M.; Pan, S.Y. The effect of high intensity ultrasonic pre-treatment on the properties of soybean protein isolate gel induced by calcium sulfate. *Food Hydrocoll.* **2013**, *32*, 303–311. [CrossRef]
34. Preece, K.E.; Hooshyar, N.; Krijgsman, A.; Fryer, P.J.; Zuidam, N.J. Intensified soy protein extraction by ultrasound. *Chem. Eng. Process. Process Intensif.* **2017**, *113*, 94–101. [CrossRef]
35. Gulseren, I.; Guzey, D.; Bruce, B.D.; Weiss, J. Structural and functional changes in ultrasonicated bovine serum albumin solutions. *Ultrason. Sonochem.* **2007**, *14*, 173–183. [CrossRef] [PubMed]
36. Xiong, W.F.; Wang, Y.T.; Zhang, C.L.; Wan, J.W.; Shah, B.R.; Pei, Y.Q.; Zhou, B.; Li, J.; Li, B. High intensity ultrasound modified ovalbumin: Structure, interface and gelation properties. *Ultrason. Sonochem.* **2016**, *31*, 302–309. [CrossRef]
37. Vera, A.; Valenzuela, M.A.; Yazdani-Pedram, M.; Tapia, C.; Abugoch, L. Conformational and physicochemical properties of quinoa proteins affected by different conditions of high-intensity ultrasound treatments. *Ultrason. Sonochem.* **2019**, *51*, 186–196. [CrossRef] [PubMed]
38. Zou, Y.; Xu, P.; Wu, H.; Zhang, M.; Sun, Z.; Sun, C.; Wang, D.; Cao, J.; Xu, W. Effects of different ultrasound power on physicochemical property and functional performance of chicken actomyosin. *Int. J. Biol. Macromol.* **2018**, *113*, 640–647. [CrossRef]
39. Burstein, E.A.; Vedenkina, N.S.; Ivkova, M.N. Fluorescence and the location of tryptophan residues in protein molecules. *Photochem. Photobiol.* **1973**, *18*, 263–279. [CrossRef]
40. Arzeni, C.; Martinez, K.; Zema, P.; Arias, A.; Perez, O.E.; Pilosof, A.M.R. Comparative study of high intensity ultrasound effects on food proteins functionality. *J. Food Eng.* **2012**, *108*, 463–472. [CrossRef]
41. Jiang, L.Z.; Wang, J.; Li, Y.; Wang, Z.J.; Liang, J.; Wang, R.; Chen, Y.; Ma, W.J.; Qi, B.K.; Zhang, M. Effects of ultrasound on the structure and physical properties of black bean protein isolates. *Food Res. Int.* **2014**, *62*, 595–601. [CrossRef]
42. Kinsella, J.E. Protein-structure and functional-properties—Emulsification and flavor binding effects. *ACS Symp. Ser.* **1982**, *206*, 301–326.
43. Peters, J.; Vergeldt, F.J.; Boom, R.M.; van der Goot, A.J. Water-binding capacity of protein-rich particles and their pellets. *Food Hydrocolloids* **2017**, *65*, 144–156. [CrossRef]
44. Sivakumar, V.; Swaminathan, G.; Rao, P.G. Use of ultrasound in soaking for improved efficiency. *J. Soc. Leather Technol. Chem.* **2004**, *88*, 249–251.

Article

Functional Properties of Rye Prolamin (Secalin) and Their Improvement by Protein Lipophilization through Capric Acid Covalent Binding

Zeinab Qazanfarzadeh [1,2], Mahdi Kadivar [1], Hajar Shekarchizadeh [1] and Raffaele Porta [2,*]

[1] Department of Food Science and Technology, College of Agriculture, Isfahan University of Technology, Isfahan 84156-83111, Iran; z.qazanfarzadeh@yahoo.com (Z.Q.); kadivar@cc.iut.ac.ir (M.K.); shekarchizadeh@cc.iut.ac.ir (H.S.)

[2] Department of Chemical Sciences, University of Naples "Federico II", Complesso Universitario di Monte Sant'Angelo, 80126 Naples, Italy

* Correspondence: raffaele.porta@unina.it; Tel.: +39-081-2539473

Abstract: Secalin (SCL), the prolamin fraction of rye protein, was chemically lipophilized using acylation reaction by treatment with different amounts of capric acid chloride (0, 2, 4, and 6 mmol/g) to enhance its functional properties. It was shown that SCL lipophilization increased the surface hydrophobicity and the hydrophobic interactions, leading to a reduction in protein solubility and water absorption capacity and to a greater oil absorption. In addition, SCL both emulsifying capacity and stability were improved when the protein was treated with low amount of capric acid chloride. Finally, the foaming capacity of SCL markedly increased after its treatment with increasing concentrations of the acylating agent, even though the foam of the modified protein was found to be more stable at the lower level of protein acylation. Technological application of lipophilized SCL as a protein additive in food preparations is suggested.

Keywords: secalin; rye prolamin; protein acylation; capric acid; emulsifying agent; foaming agent

Citation: Qazanfarzadeh, Z.; Kadivar, M.; Shekarchizadeh, H.; Porta, R. Functional Properties of Rye Prolamin (Secalin) and Their Improvement by Protein Lipophilization through Capric Acid Covalent Binding. *Foods* **2021**, *10*, 515. https://doi.org/10.3390/foods10030515

Academic Editors: Ute Schweiggert-Weisz and Emanuele Zannini

Received: 9 December 2020
Accepted: 23 February 2021
Published: 1 March 2021

1. Introduction

Wheat (*Triticum aestivum* L.) and rye (*Secale cereale* L.), the most commonly used grains in bread production [1], are closely related in taxonomy and, accordingly, their kernel endosperms contain homologous storage proteins [2,3]. However, prolamin occurring in rye grains, and alcohol-soluble protein fraction called secalin (SCL), has not been so widely studied up to now, in relation to its functionality in food systems, as wheat prolamin has been. According to previous studies, the electrophoretic pattern of SCL shows four groups of polypeptides, indicated as high molecular weight (HMW, >100 KDa), γ-75 KDa, ω (50 KDa), and γ-40 KDa proteins, respectively [4]. Moreover, quantitative amino acid analysis evinced that glutamic acid and proline are the predominant components in SCL, followed by leucine, phenylalanine, and serine [4]. Because of the SCL amphiphilic behavior due to the presence of wide regions of protein β-sheets, ethanol (70% v/v) is the most common solvent used for its extraction. In addition, ethanol leads to the unfolding of the α-helical structures increasing the extent of the β sheets and, consequently, exposing a greater number of hydrophilic groups of the protein. It is worthy to note that the β-sheet structural feature could influence SCL functional properties, i.e., those conferred by the protein to a food product [2–6] as the protein hydrophobic behavior and hence its interactions with the aqueous phase are important factors for the protein functionality. Furthermore, it is well known that protein secondary and tertiary structure may change following a chemical modification able to cause alterations in the surface exposure of their amino acids [7]. Therefore, an induced increase in protein hydrophobicity can enable its integration into lipid systems and, consequently, trigger new, as well as ameliorated or impaired, functional properties [6,8].

Protein lipophilization reaction is defined as a chemical or enzymatic structural change of a polypeptide chain by the addition of lipid components, carried out with several different methods and resulting in biomacromolecules with an increased affinity towards non-polar compounds of low or high molecular mass. Acylation by N-hydroxysuccinimide ester, acetic, succinic, or citraconic anhydrides [9–13], fatty acid chlorides [6] or reductive alkylation [14], as well as by attaching hydrophobic amino acids or alcohols [15–18], are the most frequently used experimental procedures to obtain protein lipophilization. In particular, the chemical acylation of a protein is a quite considerable structural modification that can dramatically influence the ability of the protein to interact with other molecules and, consequently, modify its activity.

The main purpose of this study was to synthesize SCL derivatives by chemical acylation with capric acid chloride (CAC), an organic compound known as a medium-chain fatty acid, and to investigate the consequent changes in SCL solubility, hydrophobicity, water, and fat binding capacity, as well as in its emulsifying and foaming properties.

2. Materials and Methods

2.1. Materials

SCL was extracted from rye flour (1:5, *w*/*v*) by 70% (*w*/*w*) ethanol containing 0.5% (*w*/*w*) sodium metabisulfite at 50 °C for 1 h and the presence of its four main polypeptide fractions (HMW, γ-75 KDa, ω, and γ-45 KDa) was then confirmed by sodium dodecyl sulfate-polyacrylamide gel electrophoresis (SDS-PAGE) under reducing conditions [4]. The composition of the SCL preparation was the following: ~91% protein (dw, N factor 5.7), ~5.2% carbohydrate (dw), ~5% moisture, ~1% lipid (dw), ~0.3% ash (dw). Corn oil was purchased from a local supermarket, whereas CAC (98% purity) and the fluorescent probe 8-anilino-1-naphthalene sulfonic acid (ANS) were purchased from Merck KGaA Co. (Darmstadt, Germany). All other chemicals and reagents utilized in this study were of analytical grade.

2.2. SCL Chemical Lipophilization

Chemical lipophilization was carried out following Schotten–Baumann's reaction. SCL was dispersed in 0.5 N NaOH (10%, *w*/*v*) and treated at both 25 and 40 °C for 15 min under stirring. Then, the pH was adjusted to 9.0 with 6 N HCl, CAC (2, 4, and 6 mmol/g of SCL) was added dropwise from a stock solution (4.7 M) to the reaction mixture under stirring, and the pH of the mixture was maintained at a value between 8.5 and 9.0 (using a pH meter Model 3310, Jenway, Staffordshire, UK) by 2 N NaOH. The reaction was considered to be completed when no more change in pH was observed. The modified SCL was then precipitated by pH adjustment to 5 using 6 N HCl, and the pellet obtained by centrifugation at 5000× *g* for 10 min was washed and centrifuged again to eliminate the excess of salts. The precipitate was finally freeze-dried, and the unreacted fatty acid present in the lipophilized SCL was eliminated by hexane extraction.

The preparation of acylated SCL tested to analyze the functional properties of the lipophilized protein was the same previously used to obtain biodegradable films [19]. The acylation efficiency and the protein recovery rate were calculated by measuring the weight and the total nitrogen content of both unmodified and lipophilized proteins [19]. Fourier Transform Infrared (FTIR) spectroscopy analysis (Model Tensor 27, Bruker Optik GmbH Co., Ettlingen, Germany) was also performed to confirm the interaction between SCL and CAC [19]. FTIR spectra of the samples were recorded within the wavenumber range of 400–4000 cm^{-1} at a resolution of 4 cm^{-1} and 32 spectra/scan at a scan speed of 2 mm/s. Lipophilization degree was determined by the elementary analysis using an Elementar Vario EL system (Model Vario EL, Elementar Analysensysteme GmbH Co., Langenselbold, Germany) and calculated on the base of the determined carbon amount. In fact, the increased carbon value compared to the control sample was converted to the interacted CAC amount value or lipophilization degree.

2.3. SCL Functional Properties

2.3.1. Surface Hydrophobicity

The surface hydrophobicity index (SHI) of both unmodified and lipophilized SCL was determined using ANS as previously described by Wan et al. [20]. Protein was dispersed (1 mg/mL) in 10 mM phosphate buffer, pH 7.0, and then diluted with the same buffer to obtain different samples with protein concentrations ranging from 0.1 to 0.25 mg/mL. Finally, 4 mL of each diluted sample were mixed with 20 μL of 10 mM phosphate buffer, pH 7.0, containing 8 mM ANS and the fluorescence intensity was measured at λ excitation = 390 nm and λ emission = 470 nm. The SHI was calculated by measuring the slope of the fluorescence intensity vs. the protein concentration.

2.3.2. Solubility Index

Protein solubility index (SI) was determined by a previously described method with slight modifications [21]. Different samples were prepared by suspending both unmodified and lipophilized SCL in distilled water (1.0% *w*/*v*) brought at different pH values (3–11) by 0.5 N HCl or NaOH. The suspensions were stirred for 30 min and their pH was checked again and readjusted to the desired value. The protein content of the supernatants obtained after centrifugation at 10,000× *g* for 20 min was determined by the Bradford assay method. In addition, protein solubility was also analyzed at pH 7 in the presence of two different NaCl concentrations (0.35 and 0.7 M). The SI was calculated by the following equation:

$$\text{SI (\%)} = (\text{soluble protein} \times 100)/\text{initial protein} \quad (1)$$

2.3.3. Emulsifying Properties

Emulsifying activity index (EAI) and emulsion stability index (ESI) were determined according to the method of Pearce and Kinsella [22] with the following modifications: 7.5 mL of either unmodified or lipophilized SCL in 0.1 M phosphate buffer, pH 7, (1.0% *w*/*v*) and 2.5 mL of corn oil were homogenized for 1 min at 14,000 rpm using an IKA ultra-turrax (Model 3725001, IKA®-Werke GmbH & Co., Deutschland, Germany) and, then, 50 μL of the prepared emulsion was dispersed into 10 mL of phosphate buffer containing 0.1% SDS (*w*/*v*). The absorbance at 500 nm of the diluted sample was measured at time 0 (A_0) and after 10 min (A_{10}). The following equations were used to calculate EAI and ESI:

$$\text{EAI (m}^2\text{/g)} = (2 \times 2.303 \times A_0 \times D)/(C \times \varnothing \times L \times 10000) \quad (2)$$

$$\text{ESI (min)} = (A_0 \times 10)/(A_0 - A_{10}) \quad (3)$$

where D is the dilution factor (200), C is the initial concentration of protein (g/mL), ∅ is the volume fraction of the oil and L is the cuvette path length (m).

2.3.4. Foaming Properties

Foaming capacity (FC) and foam stability (FS) were determined according to the method of Lawhon et al. [23]. Either unmodified or lipophilized SCL samples (1.0% *w*/*v*) were prepared in 0.1 M phosphate buffer at pH 7, homogenized for 1 min at 14,000 rpm using an IKA ultra-turrax and, then, immediately transferred into a 100 mL graduated cylinder. The total volume of foam layers was recorded at 0, 20, 40, and 60 min storage at room temperature. The following equations were used to calculate FC and FS:

$$\text{FC\%} = (\text{foam V after homogenization} \times 100)/\text{initial V of sample before homogenization} \quad (4)$$

$$\text{FS\%} = [\text{foam V after storage (at 20, 40, or 60 min)} \times 100]/\text{initial foam V} \quad (5)$$

2.3.5. Water and Oil Absorption Capacity

Oil absorption capacity (OAC) and water absorption capacity (WAC) were determined according to the method described by Beuchat [24]. Either unmodified or lipophilized SCL samples (500 mg) were dispersed in 5 mL of either distilled water or corn oil and, after stirring by vortex, were left to stand for 30 min. The samples were then centrifuged at 3000× *g* for 10 min and the volume of the supernatants were finally measured. OAC and WAC were expressed as mL of oil or water, respectively absorbed, per g of protein.

2.4. Statistical Analysis

Experiments were always carried out in triplicate. Data were processed by Excel 2010 and analysis of variance (ANOVA) was done using the Statistical Package for the Social Sciences (Version 19, SPSS Inc., Chicago, IL, USA) software to determine the significant difference between treatments. Tukey test was used to compare the mean at a 95% confidence level.

3. Results and Discussion

Figure 1 shows that the extracted SCL exhibited the well-known protein electrophoretic profile containing four fractions (HMW, γ-75 KDa, ω, and γ-45 KDa) [4].

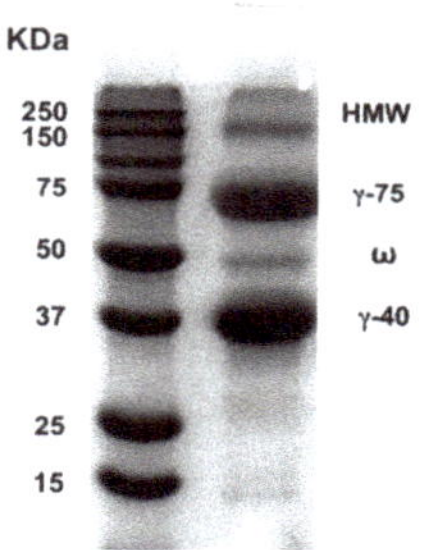

Figure 1. SDS-PAGE pattern of extracted secalin.

It is well known that all nucleophilic groups of protein amino acid residues, such as amino, hydroxyl, sulfhydryl, and phenol groups, can be acylated by fatty acid chlorides [25]. Therefore, the lateral chains of the endoprotein amino acids Lys, His, Ser, Thr, Cys, and Tyr could participate in the protein lipophilization reactions forming new O-ester, S-ester, or amide bonds. According to the previous study on the composition and amount (%) of SCL amino acids [4], the rye prolamin was shown to contain about 18% mol Lys/His/Ser/Thr/Cys/Tyr. Although some of the available hydroxyl and amino groups of these amino acids might theoretically be buried inside the protein due to its conformation and, thus, might not be able to participate in the acylation reaction, it was recently reported the SCL ability to be acylated by CAC [19]. This organic compound belongs to the class of the medium-length chain fatty acids and, thus, it is potentially able to react with polypeptides more effectively than longer fatty acids such as myristic or palmitic acids. Moreover, it has been shown that the amount of the acylated SCL increased by increasing CAC content in the reaction mixture [19]. Lipophilization degree, calculated by elementary analysis of the carbon amount occurring in the modified SCL, reached the value of ~4.6 mmol CAC/g of protein by using 6 mmol/g of CAC in the reaction mixture and, at this concentration of the acylating agent, the protein recovery rate and the acylation efficiency were ~91 and ~85%, respectively [19]. These results were also confirmed by FTIR analysis showing the formation of additional ester and amide groups in the SCL samples previously treated with CAC [19] and were consistent with the data reported by Shi et al. [26] on the lipophilization of zein with lauroyl chloride. Therefore, we were stimulated to investigate the effect of SCL treatment with different amounts of CAC (0, 2, 4, and 6 mmol/g protein) on the protein functional properties.

Since protein surface hydrophobicity is an important factor affecting its functional properties [27], the ability of both unmodified and lipophilized SCL to bind the ANS fluorescent probe was preliminarily investigated. Figure 2 shows that SCL surface hydrophobicity markedly increased as a consequence of protein lipophilization, probably because an increased number of sites accessible to ANS became available following the covalent CAC binding to the protein. The increasing of protein hydrophobicity by acylation reaction has also been reported both by Shilpashree et al. [28] and Schwenke et al. [29] following succinylation of caseins and globulins, respectively.

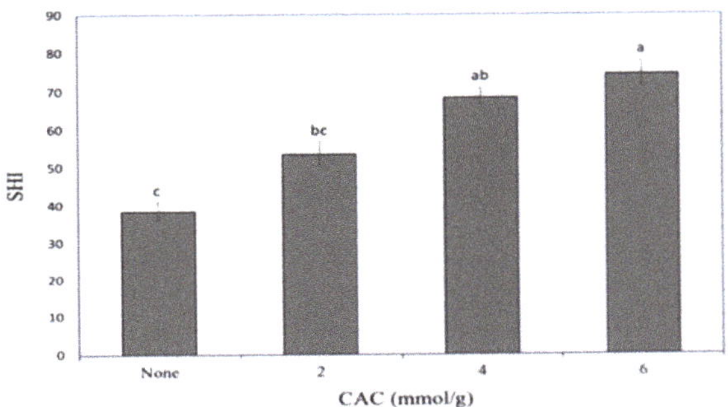

Figure 2. Effect of increasing capric acid chloride (CAC) amounts on the secalin surface hydrophobicity index (SHI). The lowercase letters (a–c) indicate significant differences among the values reported in each bar ($p < 0.05$). Further experimental details are given in the text.

The solubility of food proteins is an important property for their application in food processing because it can affect both emulsifying and foaming properties [30]. Figure 3A shows the SI of SCL, both unmodified and lipophilized by different CAC amounts, at different pH values.

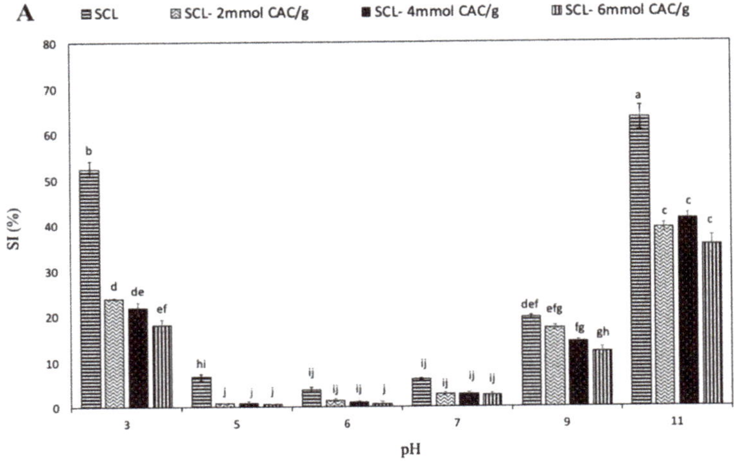

Figure 3. *Cont.*

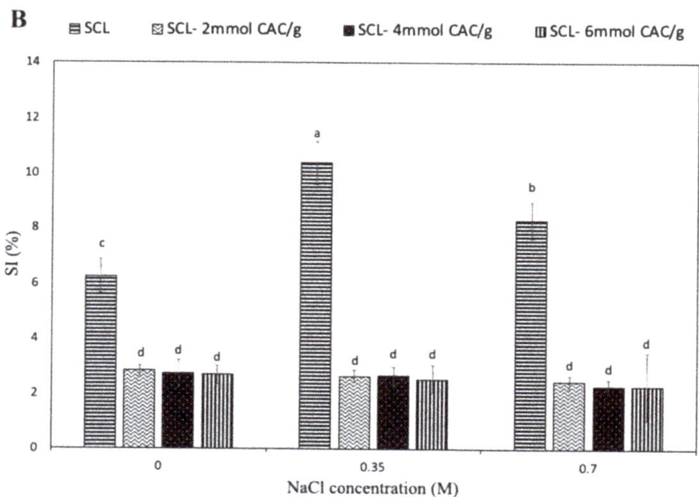

Figure 3. Effect of pH (**A**) and NaCl concentration (**B**) on the solubility index (SI) of secalin (SCL) treated or not with different amounts of capric acid chloride (CAC). The lowercase letters (a–j) indicate significant differences among the values reported in each bar ($p < 0.05$). Further experimental details are given in the text.

The results revealed that the unmodified SCL had the lowest solubility at pH 6 and that the protein solubility increased at pH values lower and higher than pH 6, at which SCL exhibited charges of higher intensity, positive and negative, respectively. A similar solubility profile was observed analyzing lipophilized SCL, even though the incorporation of CAC led to a significant reduction in protein solubility, particularly evident at acidic pH values. Such observed SI decrease could be primarily due to the increased hydrophobic features of acylated SCL related to the covalent incorporation of numerous fatty acid chains, determining a reduction in charged chemical groups and the consequent possible formation of additional hydrogen bonds. Similar results have been reported by Mendoza-Sanchez et al. [8] for bovine α-lactalbumin acylation. However, the highest SI was observed at alkaline pH values, both for unmodified and lipophilized SCL. Therefore, it is conceivable to assume that the lipophilization shifted the isoelectric point of SCL to a lower pH value, most probably because of the reduction of free amino groups after reaction with the fatty acid. Similar effects of acylation were reported in mung bean [31] and soy proteins [32]. Figure 3B shows the effect of high NaCl concentrations on SCL solubility. The SI of unmodified protein was observed to significantly increase when the saline concentration increased up to 0.35 M, probably as a consequence of a salting-in effect due to the small increase of NaCl. In fact, further enhancement of salt concentration resulted in an evident decrease in SCL solubility related to a salting-out opposite effect. Moreover, the interaction of negatively charged chloride ions with positively charged protein molecules led to a decrease in the electrostatic repulsion, enhancing the hydrophobic interactions and the consequent protein aggregation [33]. Conversely, the solubility of the lipophilized SCL seems to be not affected by the presence of high NaCl concentrations, most probably because the surface charges of SCL significantly changed with lipophilization and, consequently, NaCl was unable to influence the solubility of the modified protein.

The effects of SCL lipophilization on its WAC and OAC, as well as on the EAI and ESI, are reported in Table 1. WAC decreased in parallel with the lipophilization increase determined by the enhancing of CAC amounts in the SCL acylation reaction mixture. The observed WAC decrease is most probably due to the enlargement of the protein hydrophobic regions causing a reduced SCL ability to bind and retain water. Conversely, protein OAC markedly increased after SCL lipophilization. Furthermore, SCL EAI was found to significantly increase when 2 and 4 mmol/g of CAC were present in the acylation reaction mixture, while it significantly decreased at CAC concentrations of 6 mmol/g of SCL.

Table 1. Effect of secalin (SCL) lipophilization with increasing capric acid chloride (CAC) on protein water (WAC) and oil absorption capacities (OAC) and on its emulsifying activity (EAI) and emulsion stability index (ESI) *.

SCL	WAC (mL/g)	OAC (mL/g)	EAI (m^2/g)	ESI (min)
Unmodified	1.23 ± 0.04 ^{a}	1.05 ± 0.07 ^{c}	77.93 ± 1.82 ^{c}	21.04 ± 0.64 ^{c}
Acylated (2 mmol/g CAC)	0.83 ± 0.04 ^{b}	1.92 ± 0.10 ^{b}	121.41 ± 0.78 ^{a}	62.66 ± 1.41 ^{a}
Acylated (4 mmol/g CAC)	0.62 ± 0.03 ^{c}	2.35 ± 0.06 ^{a}	87.14 ± 2.34 ^{b}	57.68 ± 1.55 ^{a}
Acylated (6 mmol/g CAC)	0.39 ± 0.05 ^{d}	2.48 ± 0.02 ^{a}	69.27 ± 1.56 ^{d}	43.37 ± 1.33 ^{b}

* The lowercase letters (a–d) indicate significant differences among the values reported in each column ($p < 0.05$). Further experimental details are given in the text.

The ESI value of the lipophilized protein had a similar trend even though, unlike EAI, it was significantly higher than that detected with the unmodified SCL at all levels of lipophilization. It is well known that the emulsifying activity depends on the formation both of a charged layer and of an elastic film around the oil droplet, determined by the proteins located at the interfaces that cause the repulsion of the droplets, whereas the strength of the protein-protein interactions at the oil/water interfaces determines the emulsion stability. Most probably SCL lipophilization exposed the hydrophilic groups of the protein and, at the same time, oriented its hydrophobic regions towards the lipid phase, thereby increasing the emulsifying properties of SCL [34]. However, the SCL emulsifying properties were observed to decrease at the higher level of lipophilization, probably because the hydrophilic groups available for the orientation towards the aqueous phase at the oil-water interface decreased excessively in the protein acylated with higher CAC amounts and, consequently, the formation of an elastic protein film at the interface might be hindered [25,35]. Similar results were reported by Akita and Nakai [9] in their studies on the lipophilization of β-lactoglobulin by stearic acid.

Finally, the effects of lipophilization on FC of SCL and FS at pH 7 are shown in Table 2.

Table 2. Effect of secalin (SCL) lipophilization with increasing capric acid chloride (CAC) on protein foaming capacity (FC) and foam stability (FS) *.

SCL	FC (%)	FS (%)		
		20 min	40 min	60 min
SCL	53.93 ± 3.23 ^{d}	77.69 ± 3.81 ^{a}	63.51 ± 1.08 ^{ab}	39.33 ± 2.59 ^{c}
Acylated (2 mmol CAC)	160.21 ± 2.82 ^{c}	80.03 ± 4.06 ^{a}	68.12 ± 0.56 ^{a}	60.65 ± 2.46 ^{a}
Acylated (4 mmol CAC)	199.53 ± 3.53 ^{b}	76.19 ± 0.64 ^{a}	64.18 ± 3.26 ^{ab}	54.11 ± 2.58 ^{b}
Acylated (6 mmol CAC)	248.45 ± 8.48 ^{a}	69.77 ± 1.25 ^{a}	60.28 ± 0.68 ^{b}	50.98 ± 1.39 ^{b}

* The lowercase letters (a–d) indicate significant differences among the values reported in each column ($p < 0.05$). Further experimental details are given in the text.

It is well known that the FC depends on the ability of a protein to reduce the interfacial tension at the air-solution interface, which is usually accomplished by the unfolding and aligning of the proteins between the two phases [36]. Unmodified SCL was found to resist adsorption and unfolding at the air interface probably owing to its compact structure and, consequently, the formation of a suitable film was prevented. Conversely, the FC of SCL markedly increased by increasing its lipophilization with higher CAC amounts, clearly showing an improvement in the protein ability to trap air bubbles. Moreover, the FS, a measure of the ability of the foam to keep its maximum volume for a specific period [36], was also investigated for both unmodified and lipophilized SCL. The foams produced with lipophilized SCL were more stable than those obtained with unmodified SCL, so that ~61, 54, and 51% of the initial volume of foams obtained with SCL treated with 2, 4, and 6 mmol CAC, respectively, were still preserved after one-hour storage. These findings, together with those indicating an increased surface hydrophobicity and a reduced water solubility of the acylated SCL, suggest that protein-protein hydrophobic interactions are of crucial relevance for achieving stable films around air bubbles [36]. The decrease in the stability of the foams obtained with SCL lipophilized with higher CAC amounts (6 mmol/g), with respect to those obtained with SCL lipophilized with 2 mmol/g CAC, could be explained with a possible wide neutralization of the ε-amino groups of SCL lysine residues occurring following the acylation reaction. In fact, it is well known that the amino groups showed a higher tendency to participate in acylation reactions, due to the less steric hindrance, compared to other nucleophilic groups [25]. Therefore, the derived marked increase of the net negative charge of extensively acylated SCL might hinder the protein–protein interactions to form a continuous network around the air bubbles. In conclusion, the hydrophobic interactions would be predominant in SCL lipophilized with lower CAC amounts, while electrostatic repulsion would play a significant role in lowering the stability of foams obtained with SCL lipophilized at higher CAC concentrations.

Several acylation procedures carried out to modify the structure of different proteins of food interest have been previously developed in the attempt to improve their functional properties. Table 3 reports the most recent results to be compared with the present findings obtained with SCL.

A water solubility decrease similar to that reported in the present study was obtained following lipophilization of α-lactalbumin with fatty acid chloride chains longer than CAC [8], even though SCL lipophilization by CAC was found to be more effective. The improvement in SCL emulsifying properties after lipophilization was comparable to the results of other studies so that EAI value observed with SCL acylated with low CAC amounts (2 mmol/g) was almost similar to those observed using both soy protein-7S acylated with caproic acid [39] and rapeseed proteins acylated with maleic anhydride [41]. Moreover, ESI values of lipophilized SCL were close to those reported for pea and wheat gluten proteins acylated using succinic anhydride [37,38] and higher than those reported for acylated soy protein-7S, whey protein microgels, and rapeseed proteins [39–41]. On the other hand, the foaming capacity improvement of lipophilized SCL was very impressive when compared to all other previous results, even though the stability of the produced foams was similar to that of α-lactalbumin lipophilized with either lauroyl, palmitoyl, or stearoyl chlorides [8] as well as to that of whey protein microgels acetylated by acetic anhydride [40]. Conversely, WAC values of lipophilized SCL, unlike the most reported acylated proteins, showed a decreasing trend, while the increase observed in OAC values was similar to that reported for almost all the other acylated proteins. Finally, Matemu et al. [39] showed that soy protein-7S acylation by medium-, and mostly long-chain, fatty acids was able to increase the SHI more than protein acylation by short-chain fatty acids, these results being similar to those obtained by acylating SCL with high amounts of CAC.

Table 3. Effects of lipophilization of different proteins of food interest on their functional properties *.

Protein	Acylation Agent	Solubility	EAI	ESI	FC	FS	WAC	OAC	SHI
Secalin	CAC 2 mmol/g	**SD** (69%)	**SI** (56%)	**SI** (198%)	**SI** (167%)	**SI** (67%)	**SD** (33%)	(**SI** 83%)	**SI** (39%)
	CAC 4 mmol/g	**SD** (67%)	**SI** (12%)	**SI** (174%)	**SI** (270%)	**SI** (38%)	**SI** (50%)	**SI** (124%)	**SI** (78%)
	CAC 6 mmol/g	**SD** (66%)	**SD** (11%)	**SI** (106%)	**SI** (315%)	**SI** (30%)	**SD** (68%)	**SI** (136%)	**SI** (93%)
Pea proteins [37]	Succinic anhydride 0.03 g/g	**SI** (204%)	**NR**	**SI** (233%)	**SI** (64%)	**SI** (200%)	**SI** (107%)	**SD** (19%)	**NR**
Wheat gluten proteins [38]	Succinic anhydride 1.0 g/g	**SI** (583%)	**SI** (167%)	**SI** (112%)	**SI** (267%)	**SI** (457%)	**NSC**	**NR**	**NR**
Mung bean proteins [31]	Succinic anhydride 0.1 g/g	**NSC**	**SI** (42%)	**NSC**	**NR**	**NR**	**NR**	**NR**	**SD** (20%)
Soy protein-7S [39]	Caproic acid	**NR**	**SI** (50%)	**SI** (18%)	**NR**	**NR**	**SI** (192%)	**SI** (65%)	**SI** (47%)
	Caprylic acid	**NR**	**SI** (76%)	**SI** (27%)	**NR**	**NR**	**SI** (844%)	**SI** (35%)	**SI** (41%)
	Capric acid	**NR**	**SI** (86%)	**SI** (41%)	**NR**	**NR**	**SI** (760%)	**SI** (54%)	**SI** (168%)
	Lauric acid	**NR**	**SI** (100%)	**SI** (60%)	**NR**	**NR**	**SI** (388%)	**SI** (76%)	**SI** (154%)
	Myristic acid	**NR**	**SI** (70%)	**SI** (50%)	**NR**	**NR**	**SI** (745%)	**SI** (90%)	**SI** (138%)
	Palmitic acid	**NR**	**SI** (72%)	**SI** (27%)	**NR**	**NR**	**SI** (423%)	**SI** (44%)	**SI** (175%)
	Stearic acid (10.5 µmol/g)	**NR**	**SI** (70%)	**SI** (36%)	**NR**	**NR**	**NSC**	**SI** (118%)	**SI** (112%)
α-lactalbumin [8]	Lauroyl chloride	**SD** (31%)	**SI** (23%)	**SD** (62%)	**SI** (16%)	**SI** (24%)	**NR**	**NR**	**NR**
	Palmitoyl chloride	**SD** (3%)	**NSC**	**SD** (54%)	SD (32%)	**SI** (54%)	**NR**	**NR**	**NR**
	Stearoyl chloride (0.5 g/g)	**SD** (17%)	**NSC**	**SD** (35%)	**SD** (17%)	**SI** (63%)	**NR**	**NR**	**NR**
Whey protein microgels [40]	Acetic anhydride 650 mmol/mmol	**NR**	**NSC**	**SI** (15%)	**SI** (20%)	**SI** (46%)	**SD** (23%)	**SI** (32%)	**NR**
Rapeseed proteins [41]	Maleic anhydride 0.2 g/g	**SI** (27%)	**SI** (52%)	**SI** (57%)	**NSC**	**NSC**	**NR**	**NR**	**SI** (21%)

* All the data were obtained at pH 7; FS was determined after 30 min; **SI**, significantly increased; **SD**, significantly decreased; **NSC**, not significantly changed; **NR**, not reported. The values in parentheses show the differences compared to the control samples.

4. Conclusions

Changes in functional properties of SCL by capric acid incorporation were investigated. Increased protein surface hydrophobicity, oil absorption capacity, the emulsifying and foaming capacity of the lipophilized SCL, as well as the stability of both emulsions and foams obtained, were detected. Conversely, a marked decreased solubility of the protein at different pH values and at both low and high saline concentrations, as well as of its water absorption activity, were observed following SCL lipophilization. All the observed effects might be dependent on the protein unfolding and the formation of an elastic and stable protein film by hydrophobic interactions at the interfaces consequent to the capric acid covalent binding to SCL [9,20]. These findings strongly suggest the great potential of lipophilized SCL as an emulsifying and/or foaming additive in different food products.

Author Contributions: Conceptualization, M.K. and H.S.; methodology, Z.Q.; validation, Z.Q. and R.P.; investigation, Z.Q.; resources, M.K. and H.S.; writing—original draft preparation, Z.Q. and H.S.;

writing—review and editing, R.P.; supervision, M.K., H.S., and R.P.; project administration, M.K. and H.S. All authors have read and agreed to the published version of the manuscript.

Funding: This work was supported by Italian Ministries of University and Research (PRIN: Progetti di Ricerca di Interesse Nazionale—Bando 2017—CARDoon valorization by InteGrAted bio-refiNery, CARDIGAN, COD. 2017KBTK93), and by the Department of Chemical Sciences of University of Naples Federico II as sponsor of the open access publication charges.

Institutional Review Board Statement: Not applicable.

Informed Consent Statement: Not applicable.

Data Availability Statement: Data available in a publicly accessible repository.

Conflicts of Interest: The authors declare no conflict of interest.

References

1. Bushuk, W.B. Rye production and uses worldwide. *Cereal Food World* **2001**, *46*, 70–73.
2. Gellrich, C.; Schieberle, P.; Wieser, H. Biochemical characterization and quantification of the storage protein (secalin) types in rye flour. *Cereal Chem.* **2003**, *80*, 102–109. [CrossRef]
3. Shewry, P.; Parmar, S.; Miflin, B. Extraction, separation, and polymorphism of the prolamin storage proteins (secalins) of rye. *Cereal Chem.* **1983**, *60*, 1–6.
4. Qazanfarzadeh, Z.; Kadivar, M.; Shekarchizadeh, H.; Porta, R. Rye secalin characterization and use to improve zein-based film performance. *Int. J. Food Sci. Technol.* **2021**, *56*, 742–752. [CrossRef]
5. Olivera, N.; Rouf, T.B.; Bonilla, J.C.; Carriazo, J.G.; Dianda, N.; Kokini, J.L. Effect of LAPONITE® addition on the mechanical, barrier and surface properties of novel biodegradable kafirin nanocomposite films. *J. Food Eng.* **2019**, *245*, 24–32. [CrossRef]
6. Roussel-Philippe, C.; Pina, M.; Graille, J. Chemical lipophilization of soy protein isolates and wheat gluten. *Eur. J. Lipid Sci. Technol.* **2000**, *102*, 97–101. [CrossRef]
7. Foegeding, E.A.; Davis, J.P. Food protein functionality: A comprehensive approach. *Food Hydrocoll.* **2011**, *25*, 1853–1864. [CrossRef]
8. Mendoza-Sánchez, L.G.; Jiménez-Fernández, M.; Melgar-Lalanne, G.; Gutiérrez-López, G.F.; Hernández-Arana, A.S.; Reyes-Espinosa, F.; Hernández-Sánchez, H. Chemical lipophilization of bovine α-lactalbumin with saturated fatty acyl residues: Effect on structure and functional properties. *J. Agric. Food Chem.* **2019**, *67*, 3256–3265. [CrossRef]
9. Akita, E.; Nakai, S. Lipophilization of β-lactoglobulin: Effect on hydrophobicity, conformation and surface functional properties. *J. Food Sci.* **1990**, *55*, 711–717. [CrossRef]
10. Rao, A.A.; Rao, M.N. Effect of succinylation on the oligomeric structure of glycinin. *Int. J. Pept. Prot. Res.* **1979**, *14*, 307–312. [CrossRef] [PubMed]
11. Messinger, J.; Rupnow, J.; Zeece, M.; Anderson, R. Effect of partial proteolysis and succinylation on functionality of corn germ protein isolate. *J. Food Sci.* **1987**, *52*, 1620–1624. [CrossRef]
12. Franzen, K.L.; Kinsella, J.E. Functional properties of succinylated and acetylated soy protein. *J. Agric. Food Chem.* **1976**, *24*, 788–795. [CrossRef]
13. Barber, K.J.; Warthesen, J.J. Some functional properties of acylated wheat gluten. *J. Agric. Food Chem.* **1982**, *30*, 930–934. [CrossRef]
14. Lakkis, J.; Villota, R. Effect of acylation on substructural properties of proteins: A study using fluorescence and circular dichroism. *J. Agric. Food Chem.* **1992**, *40*, 553–560. [CrossRef]
15. Haque, Z.; Matoba, T.; Kito, M. Incorporation of fatty acid into food protein: Palmitoyl soybean glycinin. *J. Agric. Food Chem.* **1982**, *30*, 481–486. [CrossRef]
16. Arai, S.; Watanabe, M. Modification of succinylated αs_I-casein with papain: Covalent attachment of l-norleucine dodecyl ester and its consequence. *Agric. Biol. Chem. Tokyo* **1980**, *44*, 1979–1981. [CrossRef]
17. Aoki, H.; Taneyama, O.; Orimo, N.; Kitagawa, I. Effect of lipophilization of soy protein on its emulsion stabilizing properties. *J. Food Sci.* **1981**, *46*, 1192–1195. [CrossRef]
18. Haque, Z.; Kito, M. Lipophilization of soybean glycinin: Covalent attachment to long chain fatty acids. *Agric. Biol. Chem. Tokyo* **1982**, *46*, 597–599. [CrossRef]
19. Qazanfarzadeh, Z.; Kadivar, M.; Shekarchizadeh, H.; Porta, R. Secalin films acylated with capric acid chloride. *Food Biosci.* **2021**, *40*, 100879. [CrossRef]
20. Wan, Y.; Liu, J.; Guo, S. Effects of succinylation on the structure and thermal aggregation of soy protein isolate. *Food Chem.* **2018**, *245*, 542–550. [CrossRef] [PubMed]
21. Yuliana, M.; Truong, C.T.; Huynh, L.H.; Ho, Q.P.; Ju, Y.-H. Isolation and characterization of protein isolated from defatted cashew nut shell: Influence of pH and NaCl on solubility and functional properties. *LWT Food Sci. Technol.* **2014**, *55*, 621–626. [CrossRef]
22. Pearce, K.N.; Kinsella, J.E. Emulsifying properties of proteins: Evaluation of a turbidimetric technique. *J. Agric. Food Chem.* **1978**, *26*, 716–723. [CrossRef]
23. Lawhon, J.; Cater, C.; Mattil, K. A comparative study of the whipping potential of an extract from several oilseed flours. *Cereal Sci. Today* **1972**, *17*, 240–294.

24. Beuchat, L.R. Functional and electrophoretic characteristics of succinylated peanut flour protein. *J. Agric. Food Chem.* **1977**, *25*, 258–261. [CrossRef]
25. Mune Mune, M.; Minka, S.; Mbome, I. Effects of increasing acylation and enzymatic hydrolysis on functional properties of bambara bean (*Vigna subterranea*) protein concentrate. *Acta Aliment. Hung.* **2013**, *43*, 574–583. [CrossRef]
26. Shi, K.; Huang, Y.; Yu, H.; Lee, T.-C.; Huang, Q. Reducing the brittleness of zein films through chemical modification. *J. Agric. Food Chem.* **2010**, *59*, 56–61. [CrossRef] [PubMed]
27. Hailing, P.J.; Walstra, P. Protein-stabilized foams and emulsions. *Crit. Rev. Food Sci.* **1981**, *15*, 155–203. [CrossRef]
28. Shilpashree, B.; Arora, S.; Chawla, P.; Vakkalagadda, R.; Sharma, A. Succinylation of sodium caseinate and its effect on physicochemical and functional properties of protein. *LWT Food Sci. Technol.* **2015**, *64*, 1270–1277. [CrossRef]
29. Schwenke, K.; Mothes, R.; Zirwer, D.; Gueguen, J.; Subirade, M. Modification of the structure of 11 S globulins from plant seeds by succinylation. In *Symposium on Food Proteins: Structure-Functionality Relationships*; VCH: Weinheim, Germany, 1993; pp. 143–153.
30. Dissanayake, M.; Vasiljevic, T. Functional properties of whey proteins affected by heat treatment and hydrodynamic high-pressure shearing. *J. Dairy Sci.* **2009**, *92*, 1387–1397. [CrossRef] [PubMed]
31. Charoensuk, D.; Brannan, R.G.; Chanasattru, W.; Chaiyasit, W. Physicochemical and emulsifying properties of mung bean protein isolate as influenced by succinylation. *Int. J. Food Prop.* **2018**, *21*, 1633–1645. [CrossRef]
32. Caillard, R.; Petit, A.; Subirade, M. Design and evaluation of succinylated soy protein tablets as delayed drug delivery systems. *Int. J. Biol. Macromol.* **2009**, *45*, 414–420. [CrossRef]
33. Wanasundara, P.; Shahidi, F. Functional properties of acylated flax protein isolates. *J. Agric. Food Chem.* **1997**, *45*, 2431–2441. [CrossRef]
34. Mirmoghtadaie, L.; Kadivar, M.; Shahedi, M. Effects of succinylation and deamidation on functional properties of oat protein isolate. *Food Chem.* **2009**, *114*, 127–131. [CrossRef]
35. Johnson, E.A.; Brekke, C. Functional properties of acylated pea protein isolates. *J. Food Sci.* **1983**, *48*, 722–725. [CrossRef]
36. Casella, M.L.; Whitaker, J.R. Enzymatically and chemically modified zein for improvement of functional properties. *J. Food Biochem.* **1990**, *14*, 453–475. [CrossRef]
37. Shah, N.N.; Umesh, K.; Singhal, R.S. Hydrophobically modified pea proteins: Synthesis, characterization and evaluation as emulsifiers in eggless cake. *J. Food Eng.* **2019**, *255*, 15–23. [CrossRef]
38. Liu, Y.; Zhang, L.; Li, Y.; Yang, Y.; Yang, F.; Wang, S. The functional properties and structural characteristics of deamidated and succinylated wheat gluten. *Int. J. Biol. Macromol.* **2018**, *109*, 417–423. [CrossRef] [PubMed]
39. Matemu, A.O.; Kayahara, H.; Murasawa, H.; Katayama, S.; Nakamura, S. Improved emulsifying properties of soy proteins by acylation with saturated fatty acids. *Food Chem.* **2011**, *124*, 596–602. [CrossRef]
40. Karbasi, M.; Askari, G.; Madadlou, A. Effects of acetyl grafting on the structural and functional properties of whey protein microgels. *Food Hydrocoll.* **2020**, 106443. [CrossRef]
41. Purkayastha, M.D.; Borah, A.K.; Saha, S.; Manhar, A.K.; Mandal, M.; Mahanta, C.L. Effect of maleylation on physicochemical and functional properties of rapeseed protein isolate. *J. Food Sci. Technol.* **2016**, *53*, 1784–1797. [CrossRef]

Article

Fermentation of Lupin Protein Hydrolysates—Effects on Their Functional Properties, Sensory Profile and the Allergenic Potential of the Major Lupin Allergen Lup an 1

Katharina Schlegel [1,2], Norbert Lidzba [3], Elke Ueberham [3], Peter Eisner [2,4,5] and Ute Schweiggert-Weisz [2,6,*]

1 Department of Chemistry and Pharmacy, Friedrich-Alexander-Universität Erlangen-Nürnberg, 91054 Erlangen, Germany; katharina.schlegel@ivv.fraunhofer.de
2 Department Food Process Development, Fraunhofer Institute for Process Engineering and Packaging (IVV), 85354 Freising, Germany; peter.eisner@ivv.fraunhofer.de
3 Department of Therapy Validation, Fraunhofer Institute for Cell Therapy and Immunology (IZI), 04103 Leipzig, Germany; norbert.lidzba@izi.fraunhofer.de (N.L.); elke.ueberham@izi.fraunhofer.de (E.U.)
4 ZIEL-Institute for Food & Health, TUM School of Life Sciences Weihenstephan, Technical University of Munich, 85354 Freising, Germany
5 Faculty of Technology and Engineering, Steinbeis-Hochschule, 01069 Dresden, Germany
6 Institute of Nutritional and Food Sciences, University of Bonn, 53115 Bonn, Germany
* Correspondence: ute.weisz@ivv.fraunhofer.de; Tel.: +49-8161-491-483

Abstract: Lupin protein isolate was treated using the combination of enzymatic hydrolysis (Papain, Alcalase 2.4 L and Pepsin) and lactic acid fermentation (*Lactobacillus sakei* ssp. *carnosus*, *Lactobacillus amylolyticus* and *Lactobacillus helveticus*) to investigate the effect on functional properties, sensory profile and protein integrity. The results showed increased foaming activity (2466–3481%) and solubility at pH 4.0 (19.7–36.7%) of all fermented hydrolysates compared to the untreated lupin protein isolate with 1613% of foaming activity and a solubility of 7.3 (pH 4.0). Results of the SDS-PAGE and Bead-Assay showed that the combination of enzymatic hydrolysis and fermentation of LPI was effective in reducing *L. angustifolius* major allergen Lup an 1 to a residual level of <0.5%. The combination of enzymatic hydrolysis and fermentation enables the production of food ingredients with good functional properties in terms of protein solubility and foam formation, with a balanced aroma and taste profile.

Keywords: enzymatic hydrolysis; fermentation; lupin protein; functional properties; sensory profile; lupin allergy; lup an 1; plant protein

Citation: Schlegel, K.; Lidzba, N.; Ueberham, E.; Eisner, P.; Schweiggert-Weisz, U. Fermentation of Lupin Protein Hydrolysates—Effects on Their Functional Properties, Sensory Profile and the Allergenic Potential of the Major Lupin Allergen Lup an 1. *Foods* **2021**, *10*, 281. https://doi.org/10.3390/foods10020281

Academic Editor: Maria Antonietta Ciardiello

Received: 16 December 2020
Accepted: 29 January 2021
Published: 31 January 2021

1. Introduction

The global population is expected to grow by 2 billion to 9.7 billion people over next 30 years [1]. In order to maintain the world's population with protein, food systems will be faced with a major challenge. An increase in animal production would not be a sustainable option to meet the high demand of protein. Supplying the world with animal protein has drastic effects on the environment, which include the intensive use of land, the deterioration of air and water quality and the emission of greenhouse gases. [2]. A promising way to reduce the impact of nutrition on the environment could be the partial replacement of animal proteins with plant proteins. Legumes such as lupins are becoming more and more popular as an alternative source of protein to animal protein and soy. Lupins are widely cultivated in Europe and in South America and are particularly attractive for human consumption because of their high protein content of 39%, up to 55% (dry matter) [3] with a well-balanced amino acid profile and a low carbohydrate content compared to other legumes [4]. However, the use of lupin proteins in some foods like refreshing drinks is limited due to the characteristics of their functional properties, in particular their solubility in the acidic range. Several studies have shown that enzymatic hydrolysis can significantly improve the functional properties of plant proteins such as protein solubility, foaming and emulsification [5–10]. In addition, it was shown that the allergenicity of the proteins can be reduced by enzymatic treatments [11,12]. However, protein hydrolysis can also lead to negative modifications of the sensory profile by producing a bitter taste that inhibits their use in food [6,7,12]. One promising approach to influence the sensory profile of these ingredients could be lactic acid fermentation. Several studies have shown that fermentation of plant proteins by lactic acid leads to reduced or masked off-flavors in legumes and improves their sensory profile [13–15]. However, lactic acid fermentation is less effective in improving the functional properties of proteins and the degradation of polypeptides to reduce the allergenic potential of those proteins is less effective compared to enzymatic hydrolysis. The combination of enzymatic hydrolysis and fermentation could use the positive effects of both treatments to develop low-allergen food ingredients with excellent functional properties and a balanced sensory profile. The objective of this study was to investigate the effect of the combination of enzymatic hydrolysis and fermentation on functional properties of lupin proteins—protein solubility, foaming properties and emulsification capacity. In addition, the sensory profile of the treated ingredients was also evaluated. In order to obtain first insights of the reduction of the allergenic potential of lupin proteins, both molecular weight distribution and immunological detectability of the fermented hydrolysates were compared with untreated lupin protein isolates.

2. Materials and Methods

2.1. Raw Materials and Chemicals

2.1.1. Lupin Seeds

Lupin (*Lupinus angustifolius* L. cultivar Boregine) seeds were provided by Saatzucht Steinach GmbH & Co KG (Steinach, Germany).

2.1.2. Enzymes

The sources, types and supplier of the enzymes used in this study are listed in Table 1. Proteolytic enzyme preparations were chosen according to a previous study [7], in which promising results were achieved by those enzyme preparations in lupin protein degradation.

Table 1. Sources, types and supplier of the enzymes used in this study.

Enzyme	Type	Biological Source	Supplier
Papain	cysteine endopeptidase	Papaya (*Carica* sp.) latex	AppliChem GmbH (Darmstadt, Germany)
Alcalase 2.4 L FG	serine endopeptidase	*Bacillus licheniformis*	Novozymes A/S (Bagsvaerd, Denmark)
Pepsin	aspartic endopeptidase	Porcine (Sus domesticus) gastric mucosa	Merck KGaA(Darmstadt, Germany)

2.1.3. Strain Selection

The fermentation of lupin protein hydrolysates was carried out using *Lactobacillus sakei* ssp. *carnosus* (DSM 15831), *Lactobacillus amylolyticus* (TL 5) and *Lactobacillus helveticus* (DSM 20075). Microorganisms were purchased from Deutsche Sammlung von Mikroorganismen und Zellkulturen (Braunschweig, Germany) and Chair of Brewing and Beverage Technology (Technical University Munich, Germany). The microorganisms were stored as a cryoculture in our strain collection and were activated on MRS (De Man, Rogosa & Sharpe) agar. The selection of the microorganisms were chosen according to Schlegel, Leidigkeit, Eisner and Schweiggert-Weisz [15], based on the promising results achieved in the aroma formation and hedonic evaluation.

2.1.4. Nutrient Media

Liquid growth media and agar were obtained from Carl Roth (Karlsruhe, Germany).

2.2. Preparation of Lupin Protein Isolate

Lupin protein isolate (LPI) was prepared from *Lupinus angustifolius* L. cultivar Boregine. Lupin seeds were dehulled, separated and passed through a roller mill. The resulting flakes were de-oiled in *n*-hexane. Flakes were extracted with 0.5 M HCl (1:8 *w*/*w*) for 1 h. Suspension was separated using a decanter centrifuge at 5600× *g* and 4 °C for 1 h and the supernatant was discarded. The acid pre-extracted flakes were dispersed in 0.5 M NaOH (1:8 *w*/*w*, pH 8.0) for 1 h at room temperature while stirring and separated by centrifugation (5600× *g*, 4 °C, 1 h). The supernatant was adjusted to pH 4.5 with 0.5 M HCl. The precipitated proteins were separated by centrifugation (5600× *g*, 130 min) and neutralized with 0.5 M NaOH, pasteurized at 70 °C for 10 min) and spray-dried using an Anhydro spray dryer (SPX Flow Technology, Charlotte, NC, USA) with an inlet temperature of 180 °C and an outlet temperature of 80 °C at a mass flow rate of 24 kg/h.

2.3. Enzymatic Hydrolysis of LPI

Enzymatic hydrolysis of LPI was performed in a 5 L thermostatically controlled reaction vessel, as previously described [7]. Briefly, the protein isolate was dispersed in deionized water with an Ultra-Turrax at 5000 rpm for 1 min (IKA-Werke GmbH & Co. KG, Staufen, Germany) to achieve a protein concentration of 50 g/kg. Temperatures and pH values were adjusted, and enzyme preparation was added (Table 1). The suspension was incubated at controlled pH and temperature (Table 2) with continuous stirring for 2 h. The reaction was stopped by heating up to 90 °C for 20 min, afterwards, the suspension was cooled down to room temperature and neutralized (pH 7) with 1 M NaOH or 1 M HCl.

Table 2. Experimental design of enzymatic hydrolysis and fermentation with enzyme-to-solution ratio (E/S), temperature and pH value of enzymatic hydrolysis.

	System	E/S (%) [1]	Temperature (°C)	pH Value (-)
	Papain			
S1	*Lactobacillus sakei* ssp. *carnosus*	0.2	80	7.0
S2	*Lactobacillus amylolyticus*			
S3	*Lactobacillus helveticus*			
	Alcalase 2.4 L			
S4	*Lactobacillus sakei* ssp. *carnosus*	0.5	50	8.0
S5	*Lactobacillus amylolyticus*			
S6	*Lactobacillus helveticus*			
	Pepsin			
S7	*Lactobacillus sakei* ssp. *carnosus*	0.5	50	2.0
S8	*Lactobacillus amylolyticus*			
S9	*Lactobacillus helveticus*			

[1] enzyme-to-solution ratio of enzymatic hydrolysis.

2.4. Fermentation of Hydrolysed LPI

Hydrolysed LPI was fermented in a 5 L glass reaction vessel in an incubator under aerobic conditions and in a 5 L glass reaction vessel with a bioreactor (Biostat B, Sartorius AG, Goettingen, Germany) under anaerobic conditions, respectively, as described previously [15]. Briefly, 0.5% glucose (*w*/*w*) was added to the 5% hydrolyzed LPI (*w*/*w*) solution, pasteurized at 80 °C for 20 min and inoculated with the activated culture of 10^7 CFU/mL. Anaerobic conditions for *Lactobacillus helveticus* were achieved by flushing the reactor with N_2. LPI was fermented at 37 °C (*Lactobacillus sakei* ssp. *carnosus* and *Lactobacillus helveticus*) and 42 °C (*Lactobacillus amylolyticus*), respectively, for 24 h without stirring. Viable cell counts were determined after 0 h and 24 h of fermentation and pH course were recorded for 24 h fermentation with one measurement point at each 30 min (wtw pH 3310 pH electrode, Xylem Analytics Germany GmbH, Weilheim, Germany). The process was stopped by heating the suspension up to 90 °C for 20 min. All samples were neutralized (pH 7.0) with 1 M NaOH and spray dried with a Niro Atomizer 2238 (GEA, Düsseldorf, Germany).

2.5. Chemical Composition

Protein content was determined according to the Dumas combustion method AOAC 968.06 (TruMac N, Leco Instruments, Mönchengladbach Germany) using a protein calculation factor of N × 5.8 [16]. The dry matter (105 °C) and ash (950 °C) contents were analyzed according to AOAC methods 925.10 and 923.03 in a TGA 601 thermogravimetric system (Leco Instruments GmbH) at 105 °C and 950 °C, respectively.

2.6. Molecular Weight Distribution

The molecular weight distribution of the untreated LPI and fermented LPI hydrolysates was determined by sodium dodecyl sulfate–polyacrylamide gel electrophoresis (SDS-PAGE) as described by Laemmli [17] with modifications [15]. Briefly, LPI and fermented LPI hydrolysates were resuspended in loading buffer (0.125 mol/L Tris-HCl, 4% SDS (*w*/*v*), 20% glycerol (*v*/*v*), 0.2 mol/L DDT, 0.02% bromophenol blue, pH 6.8), dissolved in an ultrasonic bath (30 °C, 30 min), boiled at 95 °C for 5 min (Eppendorf Thermomixer, Eppendorf AG, Hamburg, Germany) and separated with a Mini Spin centrifuge at 12,045× *g* for 10 min (Eppendorf AG). Supernatant was mixed in a ratio of 1:10 with loading buffer (see above). An aliquot of 10 µL of each sample was transferred into the wells of Bio-Rad 4–20% Criterion TGX Stain-Free precast gels (Bio-Rad Laboratories GmbH, Feldkirchen, Germany). The Precision Plus Protein™ Unstained Protein Standard (Bio-Rad Laboratories) was used as molecular weight marker (10–250 kDa). Gels were run at room temperature for 38 min at 200 V (60 mA, 100 W) in a vertical electrophoresis cell (Bio-Rad Laboratories). Protein bands were visualized using a Gel Doc™ EZ Imager system (Bio-Rad Laboratories) and determined using Image Lab software (Bio-Rad Laboratories).

2.7. Determination of Lup an 1 with Specific Monoclonal Antibodies

Amount of Lup an 1 was measured by an in house sandwich assay using β-conglutin specific antibodies (Izimab Lup an 1-1 and Izimab Lup an 1-2 (unpublished)) on FLEXMAP 3D® flow analyzer system from Luminex Corporation (Austin, TX, USA). The monoclonal capture antibody, Izimab Lup an 1-1, was coupled onto MagPlex® beads and the monoclonal detection antibody Izimab Lup an 1-2 was biotinylated according to standard procedures. Lupin samples were extracted using denaturing conditions [6]. The samples were incubated with bead conjugated Izimab Lup an 1-1, washed three times in PBS with 0.05% Tween® 20 (PBS-T) and incubated with biotinylated Izimab Lup an 1-2 followed by another wash with PBS-T. Streptavidin-conjugated phycoerythrin (SA-PE) was diluted 1:2000 in assay buffer, 100 µL per well were added to the sandwich complex and agitated for one hour. The reaction was terminated by washing the wells three times with wash buffer and subsequently filled up with 120 µL assay buffer. Readings on FLEXMAP 3D® flow analyzer expressed as median fluorescence intensity (MFI) per 100 beads calculated on the basis of a calibration curve were converted to relative reactivity in percent by dividing the individual MFI by the MFI of the control sample, which represents the LPI used as raw material in the processing steps.

2.8. Technofunctional Properties

2.8.1. Protein Solubility

The solubility (%) of LPI and fermented LPI hydrolysates was determined in duplicate at pH 4.0 and 7.0 according to Morr, et al. [18]. The sample was suspended in 0.1 M NaCl (3% *w*/*w*), pH was adjusted (0.1 M HCl) and stirred for 1 h at room temperature. Undissolved fractions of the samples were removed by centrifugation (20,000× *g*, 15 min) at room temperature. The supernatants were filtered with a Whatman No. 1 filter paper. Protein content was determined by a method of Lowry, et al. [19] using the DC Protein Assay (Bio-Rad Laboratories) with a BSA standard curve for calculating the protein concentration. The absorbance was read at 750 nm. The resulting protein content was related to the total amount of protein and protein solubility (%) was determined.

2.8.2. Foaming Properties

Foaming activity and stability were determined according to Phillips, et al. [20] in duplicate. For foaming activity, a 100 mL solution of 5% protein (pH 4.0 and 7.0) was whipped at room temperature for 8 min in a whipping machine (Hobart 50-N, Hobart GmbH, Offenburg, Germany. The increase of foam volume was defined as foaming activity (%); the percentage of foam volume remaining after 1 h was defined as foam stability (%).

2.8.3. Emulsifying Capacity

Emulsifying capacity was determined at pH 4.0 and 7.0 according to Wang and Johnson [21]. Samples were suspended in deionized water (1% *w*/*w*), adjusted to pH and stirred with an Ultraturrax (IKA-Werke GmbH & Co. KG) at 18 °C. Rapeseed oil was continuously added (10 mL/min) until phase inversion was detected (<10 µS/cm). The volume of added oil was used to calculate emulsifying capacity (mL oil per g sample). Measurements were performed in duplicate.

2.9. Sensory Analysis of Fermented Hydrolysates

2.9.1. Panelists

Panelists were trained assessors recruited from Fraunhofer IVV (Freising, Germany), with no known illness and normal olfactory function during the test. The panel consisted of 10 panelists. All panelists were tested for their olfactory function in weekly training sessions with selected suprathreshold aroma solutions. The samples were evaluated during two sessions on one day.

2.9.2. Descriptive Analysis

Samples (2% *w*/*w*) of LPI and fermented LPI hydrolysates were prepared in tap water by stirring. All samples were presented to the panel in covered glass vessels. The panelists were requested to open the lid of the vessels and record the retronasal aroma and taste attributes. After a short discussion, common retronasal aroma attributes with corresponding references and taste attributes were collected and rated on a scale from 0 (no perception) to 10 (strong perception) by each panelist in a separate session. The following ten aroma and taste qualities and corresponding references (given in brackets) were selected by the trained panelists (n = 10) for LPI and fermented LPI hydrolysates: oatmeal-like (oatmeal); cocoa-like (cocoa); malty (methylpropanal); green, grassy (hexanal) pea-like (2–isopropyl-3-methoxypyrazine); fatty ((E,Z)-2,4-nonadienal); cardboard-like, cucumber-like ((E)-2-nonenal); roasty (2-acetylpyrazine); cooked potato-like (3-(methylthio)propanal); earthy (2,3- diethyl-5-methylpyrazine), bitter, salty, sour. The sensory evaluation was carried out once.

2.10. Statistical Analysis

Results are expressed as means ± standard deviations and, for sensory evaluation, (aroma profile) as median ± standard deviations. Data were analyzed using pairwise *t*-test to determine the significance of differences between a sample and the untreated LPI, with a threshold of $p < 0.05$. Statistical analysis and visualization were performed with Origin 2018 for Windows (Origin Lab Corporation, Northampton, MA, USA). The results of the sensory evaluation were evaluated using Principal Component Analysis (PCA) covariance matrix to assess aroma and taste qualities. PCA was performed using Origin 2018 for Windows (Origin Lab Corporation).

3. Results and Discussion

3.1. Chemical Properties

Dry matter content (%) of all fermented LPI hydrolysates were within the range of 92.8% for the Alcalase 2.4 L hydrolysate S4 (*Lactobacillus sakei* ssp. *carnosus*) to 94.0% for the Pepsin hydrolysate S7 (*Lactobacillus sakei* ssp. *carnosus*)—a significantly ($p < 0.05$) lower dry matter content compared to untreated LPI (95.4%) (Table 3).

Table 3. Dry matter (%), protein content (%) and ash content (%) of lupin protein isolate (LPI) and fermented LPI hydrolysates.

Samples	Dry Matter (%)	Protein Content (%)	Ash Content (%)
LPI	95.4 ± 0.0	89.6 ± 0.0	4.2 ± 0.12
Papain			
S1	93.8 ± 0.0 *	74.7 ± 2.5	6.7 ± 0.9
S2	93.6 ± 0.2 *	78.7 ± 2.0	5.4 ± 0.6
S3	93.1 ± 0.0 *	66.8 ± 0.0 *	6.5 ± 1.1
Alcalase 2.4 L			
S4	92.8 ± 0.2 *	74.8 ± 0.6 *	6.2 ± 0.9 *
S5	93.5 ± 0.3 *	75.6 ± 1.3 *	5.3 ± 0.4
S6	93.0 ± 0.2 *	77.2 ± 5.6	7.4 ± 0.1 *
Pepsin			
S7	94.0 ± 0.0 *	73.1 ± 5.9	7.1 ± 0.4
S8	93.9 ± 0.1 *	78.0 ± 1.5	6.3 ± 0.6
S9	93.7 ± 0.2 *	72.2 ± 4.5	8.8 ± 0.7 *

The data are expressed as mean ± standard deviation ($n = 4$). Means marked with an asterisk (*) within a column indicate significant differences between the sample and the untreated LPI ($p < 0.05$) following pairwise t-test.

Untreated LPI contained the highest protein content with 89.6%. Protein content of fermented LPI hydrolysates ranged from 66.8% for *Lactobacillus amylolyticus* fermented Papain hydrolysate (S3) to 78.7% for *Lactobacillus helveticus* fermented Papain hydrolysate (S2).

Ash content (%) was within the range of 4.2% for LPI to 8.8% for the Pepsin hydrolysate S9 (*Lactobacillus helveticus*). The increased ash content of treated samples compared to untreated LPI might be attributed to the addition on NaOH during neutralization after fermentation.

3.2. Comparison of Microbial Growth on Lupin Protein Isolate Solutions

The growing parameters (CFU and pH) for all experiments after 0 h and 24 h of fermentation are shown in Tables 4 and 5. The results showed that all microorganisms were able to grow in hydrolyzed LPI. The minimum increase in CFU/mL ($\Delta E_{CFU/mL}$) was recorded for Pepsin hydrolysate S9 (*Lactobacillus helveticus*) with 1.02×10^8 CFU/mL and the maximum for Pepsin hydrolysate S8 (*Lactobacillus amylolyticus*) with 1.32×10^9 CFU/mL.

Table 4. Colony forming units (CFU) after 0 h and 24 h of fermentation.

Samples	CFU/mL		
	0 h	24 h	$\Delta E_{CFU/mL}$
Papain			
S1	$2.25 \times 10^7 \pm 2.12 \times 10^6$	$4.86 \times 10^8 \pm 7.85 \times 10^7$	$4.63 \times 10^8 \pm 8.06 \times 10^7$
S2	$1.03 \times 10^7 \pm 1.20 \times 10^6$	$1.03 \times 10^9 \pm 1.41 \times 10^7$	$1.02 \times 10^9 \pm 1.29 \times 10^7$
S3	$6.45 \times 10^6 \pm 1.30 \times 10^6$	$2.10 \times 10^8 \pm 5.21 \times 10^7$	$2.03 \times 10^8 \pm 5.09 \times 10^7$
Alcalase 2.4 L			
S4	$1.26 \times 10^7 \pm 1.91 \times 10^6$	$1.30 \times 10^9 \pm 7.21 \times 10^8$	$1.29 \times 10^9 \pm 7.23 \times 10^8$
S5	$1.36 \times 10^7 \pm 1.41 \times 10^6$	$1.23 \times 10^9 \pm 3.56 \times 10^8$	$1.21 \times 10^9 \pm 3.54 \times 10^8$
S6	$8.45 \times 10^6 \pm 1.53 \times 10^6$	$3.62 \times 10^8 \pm 1.63 \times 10^8$	$3.54 \times 10^8 \pm 1.61 \times 10^8$
Pepsin			
S7	$7.01 \times 10^6 \pm 1.34 \times 10^5$	$1.97 \times 10^8 \pm 1.91 \times 10^7$	$1.89 \times 10^8 \pm 1.90 \times 10^7$
S8	$1.14 \times 10^7 \pm 2.88 \times 10^6$	$1.34 \times 10^9 \pm 2.62 \times 10^8$	$1.32 \times 10^9 \pm 2.59 \times 10^8$
S9	$1.06 \times 10^7 \pm 2.12 \times 10^6$	$1.12 \times 10^8 \pm 6.48 \times 10^7$	$1.02 \times 10^8 \pm 6.70 \times 10^7$

The data are expressed as mean ± standard deviations from duplicates.

Table 5. pH values after 0 h and 24 h of fermentation.

Samples	pH-Values	
	0 h	**24 h**
Papain		
S1	7.1 ± 0.1	4.6 ± 0.1
S2	7.1 ± 0.0	4.9 ± 0.1
S3	6.9 ± 0.3	3.3 ± 0.2
Alcalase 2.4 L		
S4	7.1 ± 0.0	5.1 ± 0.0
S5	7.0 ± 0.1	4.8 ± 0.1
S6	7.3 ± 0.3	4.9 ± 0.4
Pepsin		
S7	7.0 ± 0.0	4.9 ± 0.1
S8	7.1 ± 0.0	4.9 ± 0.0
S9	7.3 ± 0.2	3.7 ± 0.3

The data are expressed as mean ± standard deviations from duplicates.

The lowest pH values after 24 h of fermentation were recorded for Papain hydrolysate S3 (*Lactobacillus helveticus*) and Pepsin hydrolysate S9 (*Lactobacillus helveticus*) with pH values of 3.3 and 3.7, respectively. Fermentation of hydrolysates obtained by Alcalase 2.4 treatment (S4–S6) tended to show higher pH values compared to hydrolysates obtained by other proteolytic enzyme preparations. Presumably, Alcalase 2.4 degradation leads to a higher buffer capacity of the respective samples. The lowest pH reduction (5.1) was achieved by Alcalase 2.4 hydrolysate S4 (*Lactobacillus sakei* ssp. *carnosus*). A former study described a rapid decrease in pH for *Lactobacillus helveticus* and *Lactobacillus amylolyticus* during the fermentation of LPI [15]. In this work, the pH curve over 24 h of fermentation was also recorded (Figure 1 shows exemplarily the pH curve for LPI samples (not hydrolyzed) fermented with *Lactobacillus helveticus*. It was observed that *Lactobacillus helveticus* showed an extended lag phase of 7 h in decreasing the pH value than *Lactobacillus helveticus* did on Papain (S3) and Pepsin (S9) hydrolysate, respectively. However, after 24 h of fermentation, *Lactobacillus helveticus* was able to decrease the pH in the range of 3.1 and 4.6 in all hydrolysates.

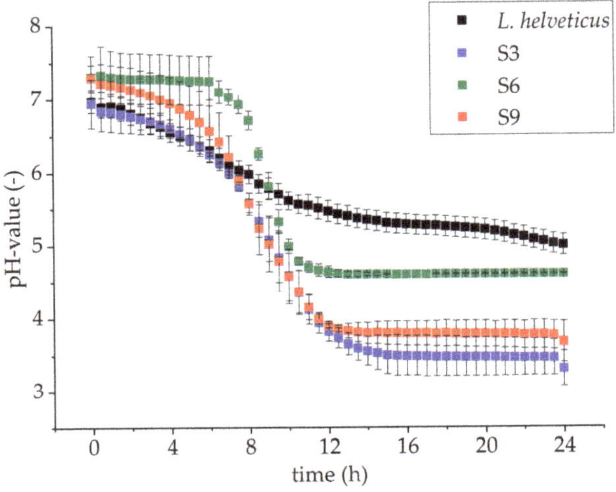

Figure 1. Course of pH value for *Lactobacillus helveticus* fermentation on LPI (black curve) and on Papain (S3, blue curve), Alcalase 2.4 L (S6, green curve) and Pepsin (S9, red curve) LPI hydrolysates over 24 h fermentation. The data are expressed as mean ± standard deviations from duplicates.

3.3. Molecular Weight Distribution (SDS-PAGE) and Immunoreactivity

The molecular weight distribution (SDS-PAGE) of LPI and its fermented hydrolysates was used to determine the protein integrity and is shown in Figure 2. All treatments resulted in prominent changes in the SDS-PAGE profile with hydrolyzed polypeptides to smaller fragments with molecular weights below 30 kDa. Enzymatic hydrolysis seems to have the greatest influence on the degradation of polypeptides of LPI. Comparing the profiles of the molecular weight distribution of a previous study [7] after hydrolysis of LPI, it is shown that the profiles did not change visibly compared to those after the combination of hydrolysis and fermentation. This observation is also supported by the results of a further study [15], which show that fermentation has only minor influence on the profiles of molecular weight distribution. Furthermore, it is described that β-conglutin with a molecular weight of ~55–61 kDa is known as the major allergen of *L. angustifolius* L. (Lup an 1) [22]. The SDS-PAGE results of all fermented LPI hydrolysates showed a degradation of the described IgE-reactive polypeptide (Figure 2) independent of the enzyme preparations used. These observations were confirmed by the results of the Bead-Assay (Figure 3). A significant reduction in the signal intensity of Lup an 1 below 0.5% was observed for all fermented hydrolysates compared to unfermented LPI (100%) due to the combination of enzymatic hydrolysis and fermentation The highest decrease of immunological reactivity was achieved by treatment with Alcalase 2.4 L combined with fermentation S5 (*Lactobacillus amylolyticus*) and S6 (*Lactobacillus. helveticus*). No binding of Lup an 1 antibodies in sandwich format could be detected (read out not detectable). The assumption that enzymatic hydrolysis is a powerful approach to reduce allergenic potential is supported by further studies [23–27].

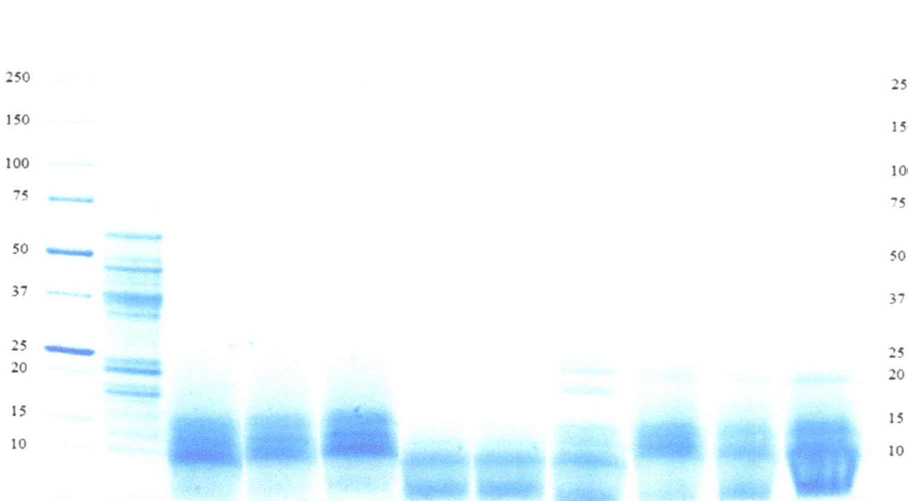

Figure 2. Peptide band profiles of LPI and fermented LPI hydrolysates as determined by SDS-PAGE under reducing conditions.

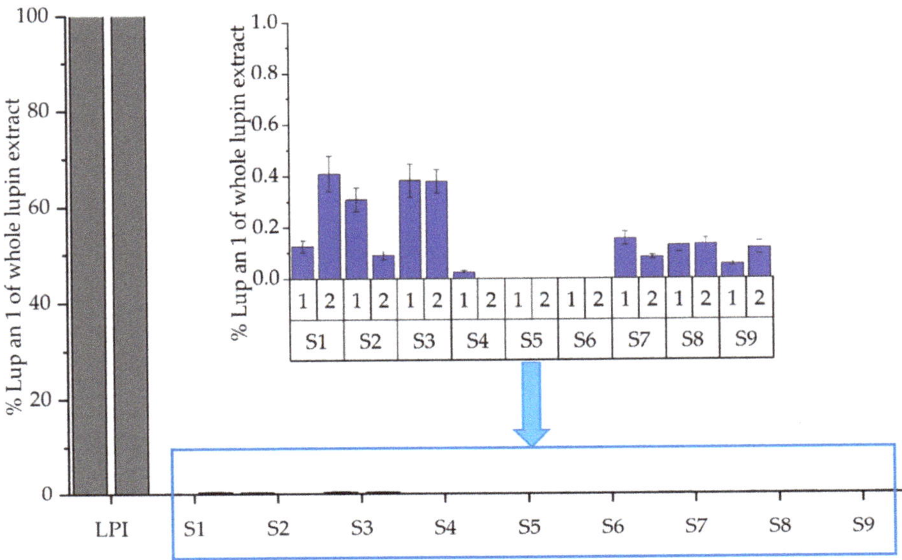

Figure 3. Determination of Lup an 1 with specific monoclonal antibodies (Izimab Lup an 1-1 and Izimab Lup an 1-2) of unfermented LPI and fermented LPI hydrolysates using Bead-Assay. Results are shown as mean ± standard derivation of each duplicate (1 and 2). Each Bead-Assay was performed in triplicate.

3.4. Technofunctional Properties

3.4.1. Protein Solubility

Protein solubility of LPI and fermented LPI hydrolysates was determined as a function of pH at pH 4.0 and pH 7.0 and is given in Table 6. All samples showed higher protein solubility at pH 7.0 than at pH 4.0. Solubility decreases as pH value approaches the isoelectric point, approximately pH 4.5–5.0, as discussed frequently [4,7,12,28,29]. Protein solubility at pH 7.0 ranged for all samples, from 45.2% for the Papain hydrolysate S2 (*Lactobacillus helveticus*) to 66.6% for Papain hydrolysate S3 (*Lactobacillus amylolyticus*). However, a significant difference between LPI and the fermented hydrolysates (with the exception of Pepsin hydrolysate S6 (*Lactobacillus helveticus*)) could not be observed.

In contrast, protein solubility at pH 4.0 after enzymatic and fermentation treatment was significantly different ($p < 0.05$) in comparison to untreated LPI (7.3%). All fermented hydrolysates showed significant higher protein solubility with values between 19.7% and 36.7%. The minimum protein solubility (19.7%) of the fermented hydrolysates was determined after fermentation of the Papain hydrolysate S2 (*Lactobacillus helveticus*). The fermentation of a nonhydrolyzed lupin protein isolate with *Lactobacillus helveticus* also resulted in very low protein solubility at pH 4.0 and 7.0 in a former study [15] The single hydrolysis of LPI by means of Alcalase 2.4 L resulted in a considerable increase in protein solubility compared to the hydrolysis by means of other proteolytic enzyme preparation [7], confirming high proteolytic activity of the Alcalase 2.4 L preparation. However, the tendency for the highest protein solubility for Alcalase 2.4 L-hydrolyzed LPI at both pH values and significantly lower protein solubility for Papain and Pepsin-hydrolyzed LPI compared to Alcalase 2.4 L hydrolysate was not observed in this study. A significant difference between the applied enzymes could not be identified.

Table 6. Protein solubility, foam properties and emulsifying capacity of LPI and fermented LPI hydrolysates.

Samples	Protein Solubility		Foaming Activity		Foam Stability		Emulsifying Capacity	
	pH 4.0	pH 7.0	pH 4.0	pH 7.0	pH 4.0	pH 7.0	pH 4.0	pH 7.0
	(%)	(%)	(%)	(%)	(%)	(%)	(%)	(%)
LPI	7.3 ± 0.3	63.6 ± 3.0	828 ± 3	1613 ± 11	92 ± 1	89 ± 0	410 ± 7	666 ± 0
Papain								
S1	24.8 ± 3.2 *	53.3 ± 7.7	2118 ± 47 *	2544 ± 39 *	0 ± 0 *	1 ± 0 *	350 ± 21	393 ± 15 *
S2	19.7 ± 2.4 *	45.2 ± 10.7	1819 ± 38 *	2606 ± 53 *	0 ± 0 *	38 ± 8 *	340 ± 0 *	432 ± 65 *
S3	36.7 ± 3.0 *	66.6 ± 6.9	2395 ± 45 *	2505 ± 54 *	0 ± 0 *	29 ± 6 *	415 ± 28	432 ± 9 *
Alcalase 2.4 L								
S4	23.4 ± 4.6 *	60.4 ± 9.3	2458 ± 58 *	2466 ± 54 *	0 ± 0 *	96 ± 3	358 ± 32	246 ± 28 *
S5	25.6 ± 8.8 *	55.0 ± 7.9	2766 ± 54 *	2676 ± 56 *	0 ± 0 *	96 ± 2 *	258 ± 18	283 ± 30 *
S6	27.3 ± 1.7 *	46.0 ± 3.8 *	2789 ± 28 *	2721 ± 91	0 ± 0 *	0 ± 0 *	429 ± 8 *	323 ± 3 *
Pepsin								
S7	25.7 ± 5.0 *	57.5 ± 12.2	1819 ± 51 *	3338 ± 71 *	0 ± 0 *	41 ± 1 *	400 ± 35	477 ± 16 *
S8	24.0 ± 2.1 *	56.8 ± 6.3	1993 ± 46 *	3481 ± 39 *	0 ± 0 *	90 ± 4	400 ± 7 *	439 ± 5 *
S9	25.6 ± 2.7 *	55.7 ± 11.9	2001 ± 40 *	3443 ± 51 *	0 ± 0 *	7 ± 1 *	405 ± 0	417 ± 6 *

The data are expressed as mean ± standard deviation ($n = 4$). Means marked with an asterisk (*) within a column indicate significant differences between sample and untreated LPI ($p < 0.05$) following pairwise t-test.

3.4.2. Foaming Properties

Foaming activity (%) and foam stability (%) of untreated LPI and fermented LPI hydrolysates are given in Table 6. All treated samples showed significantly ($p < 0.05$) higher foaming activity, with values from 1819% up to 2789% at pH 4.0 and 2466% up to 3481% at pH 7.0, compared to untreated LPI with values of 828% and 1613%, respectively. The highest activities with values of 3338%, 3443% and 3481% were determined for Pepsin hydrolysates (S7–S9) at pH 7.0. A high foaming activity of 3614% was also observed at pH 7.0 after single hydrolysis of LPI with Pepsin [7]. The foaming activities at pH 4 were lower than those at pH 7 for all samples, with the exception of the samples hydrolyzed with Alcalase 2.4 L (S4–S6). It is assumed that the lower foam activity under acidic conditions was caused by the significantly lower protein solubility of the samples at pH 4.0 in comparison to pH 7.0.

Furthermore, previous studies have shown that foaming activity of LPI increases only slightly through fermentation [15,30]. Therefore, enzymatic treatment seems to have the greatest effect on foaming activity in this study. Enzymatic hydrolysis breaks larger polypeptides into smaller peptides, thus improving foam formation by rapid diffusion at the air–water interface [31]. Furthermore, Meinlschmidt, Schweiggert-Weisz and Eisner [26] have shown an increase in foaming activity of soy protein isolates after enzymatic hydrolysis and fermentation.

LPI showed foam stability (%) of 89% at pH 7.0 after 1 h standing. The samples treated with *Lactobacillus amylolyticus* and Pepsin (S8), as well as the samples treated with *Lactobacillus sakei* ssp. *carnosus* and Alcalase 2.4 L (S4), did not show significant differences in foam stability with 90% and 96%, respectively, in comparison to the untreated LPI. All other treated samples showed significantly lower or even no foam stability. At pH 4.0, the untreated LPI showed foam stability of 92%. All fermented LPI hydrolysates did not show any foam stability. Foams can be stabilized by large peptides with flexible structures. Hydrolysis reduces protein surface coverage, which means that the air–water interface is no longer stabilized and foam collapse occurs in the hydrolyzed protein foams [32].

3.4.3. Emulsifying Capacity

The emulsifying capacity of untreated LPI at pH 7.0 was 666 mL/g. As shown in Table 6, all treated samples showed significantly ($p < 0.05$) lower emulsifying properties than LPI with values lower than 477 mL/g. The samples hydrolyzed with Alcalase 2.4 L showed the lowest emulsifying properties with values of 246 (S4), 283 (S5) and 323 mL/g (S6). The emulsifying capacity at pH 4.0 showed lower values than at pH 7.0 for all samples, with the exception of the samples treated with Alcalase 2.4 L (S4–S6). El-Adawy, Rahma, El-Bedawey and Gafar [32] found a direct correlation between the emulsifying properties and solubility of a protein. In this study, we could not find a correlation between the solubility of proteins and their ability to form an emulsion within the same pH value. However, we observed that emulsification capacities were lower at pH 4.0 than at pH 7.0. In addition, the samples at pH 4.0 showed lower protein solubilities than at pH 7.0.

3.5. Sensory Analysis

Fermentation enables changes in sensory profiling through the generation and degradation of flavor active compounds. Proteins, carbohydrates and fats from the raw materials provide the necessary precursors, e.g., for the formation of volatile aroma-active compounds. Most of the formation pathways of aroma-active and organic compounds are based on functional metabolic pathways of lactic acid bacteria.

Comparative retronasal aroma profile analysis (Figure 4) shows the results of Papain hydrolysate S2 (*Lactobacillus amylolyticus*) and Alcalase 2.4 L hydrolysate S4 (*Lactobacillus sakei* ssp. *carnosus*) compared to LPI across the panel and highlights the most impressive changes with significant differences ($p < 0.05$) in aroma perception of cocoa-like and malty. For LPI, the major aroma perceptions were rated with a moderate perception with values of 3.5 for oatmeal-like, 3.0 for fatty and 3.0 for pea-like. All other aroma impressions were rated with an intensity of 2.6 and below. The Papain hydrolysate S2 (*Lactobacillus amylolyticus*) and the Alcalase 2.4 L hydrolysate S4 (*Lactobacillus sakei* ssp. *carnosus*) were the only samples that showed significant differences ($p < 0.05$) in the intensity of aroma perception (cocoa-like and malty) compared to untreated LPI. The cocoa-like impression increased with values of 2.5 for S2 and 6.3 for S4 compared to 0.8 for unmodified LPI. Furthermore, S4 showed a significant increase ($p < 0.05$) in the intensity of the aroma perception of malty (3.7). All other samples showed no significant differences ($p < 0.05$) of aroma perceptions compared to those of unmodified LPI, and are therefore not presented in the figure (see Supplementary Materials).

The taste impressions bitter, salty and sour of LPI and fermented LPI hydrolysates were investigated by 10 panelists in the sensory evaluation. The bitter, salty and sour intensities of LPI were described with values of 3.0, 2.1 and 1.0, respectively. All treated samples did not show significant difference ($p < 0.05$) compared to untreated LPI, with the exception of the sample treated with Alcalase 2.4 L hydrolysate S5 (*Lactobacillus amylolyticus*), with a more bitter intensity of 5.6. Additionally, Pepsin hydrolysate S9 (*Lactobacillus helveticus*) showed significantly ($p < 0.05$) higher intensity of saltiness compared to untreated LPI. No significant differences could be observed between LPI and the treated samples regarding the intensity of sour.

Principal Component Analysis (PCA) was applied to identify the most relevant sensory attributes that affect sample characteristics. Figure 5 shows the resulting biplots of the uncorrelated principal components (PCs) 1 and 2 based on the sensory data of the respective hydrolysates including untreated LPI, fermented LPI hydrolysates and the scaled loadings.

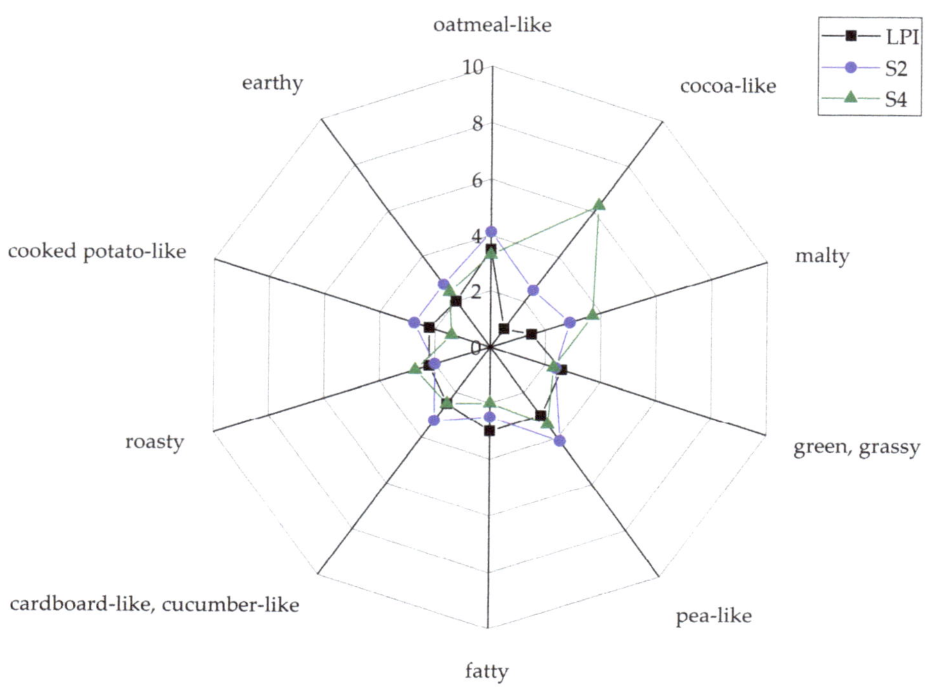

Figure 4. Comparative retronasal aroma profile analyses of LPI and fermented LPI hydrolysates (S2 and S4) on a scale from 0 (no perception) to 10 (strong perception). The data are displayed as mean values of the sensory evaluation (n = 10).

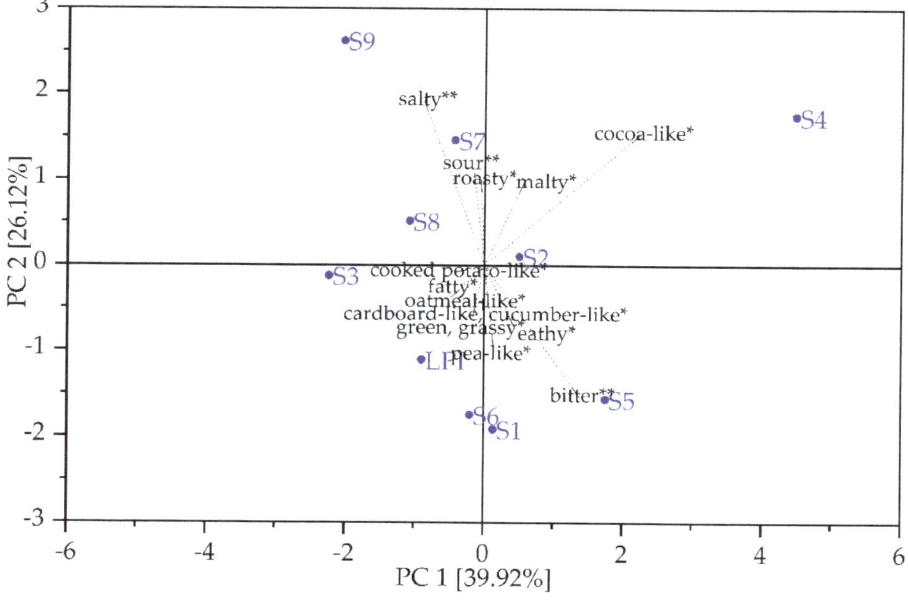

Figure 5. Biplot of aroma and taste of the unfermented LPI and fermented LPI hydrolysates. * aroma ** taste.

The first two components of PCA explained 39.92% and 26.12% of the observed variation (68.22% in total). The attribute with the strongest influence on PC1 was salty (−0.298) taste and cocoa-like (0.754) aroma impression. In contrast, PC2 was primarily described by the attributes bitter (−0.412) and salty (0.514). Unfermented LPI (−0.896/−1.108) was in close correlation with the pea-like aroma attribute and was found on the negative side of PC2, together with the Papain hydrolysates samples S1 (*Lactobacillus sakei* ssp. *carnosus*) and S3 (*Lactobacillus helveticus*) and the Alcalase 2.4 L hydrolysates samples S5 (*Lactobacillus amylolyticus*) and S6 (*Lactobacillus helveticus*). The samples S2, S4 and the Pepsin hydrolysates S7 (*Lactobacillus sakei* ssp. *carnosus*), S8 (*Lactobacillus amylolyticus*) and S9 (*Lactobacillus helveticus*) were found on the positive side and were opposite of the unfermented LPI. The Pepsin hydrolysate fermented with *L. helveticus* (S9) scored the highest in the PC2 (2.623) and was nearest with salty. S9 raised a low pH value of 3.7 after fermentation, which was neutralized (pH 7.0) with 1M NaOH. Due to the larger amount of NaOH, compared to other samples, the salty impression may have been increased. This assumption is further supported by the increased ash content for S9. In contrast, the Papain hydrolysate S1 (*Lactobacillus sakei* ssp. *carnosus*) and the Alcalase 2.4 L hydrolysates S5 (*Lactobacillus amylolyticus*) and S6 (*Lactobacillus helveticus*) scored the lowest in PC2 and were nearest ranged to the bitter taste attribute. It was observed that the Alcalase 2.4 L hydrolysate S5 (*Lactobacillus amylolyticus*) showed a significant difference ($p < 0.05$) of bitter intensity to untreated LPI. The Alcalase 2.4 L hydrolysates S4 (*Lactobacillus sakei* ssp. *carnosus*) and S6 (*Lactobacillus helveticus*) did not show a significant difference ($p < 0.05$), but a tendency of higher bitter intensity. Alcalase 2.4 L hydrolysates are known for bitter taste. Several studies have already shown that the hydrolysis of plant proteins such as lupin [7,10], soy [6] and pea [27] results in a bitter taste. However, S4 (*Lactobacillus sakei* ssp. *carnosus*) was not ranged close to the bitter attribute, rather in the positive range of PC2 (1.746), and was nearest ranged with cocoa-like and was scored the highest in the PC1 (4.483).

4. Conclusions

The combination of enzymatic hydrolysis and fermentation of lupin protein isolate can increase foaming activity while maintaining proper emulsification capacity. In addition, the modification increases protein solubility at acidic conditions and thus opens new possibilities in food applications such as refreshing drinks. LPI provides a well-balanced sensory profile that has been partially altered by the treatments in aroma and taste perception. The sensory acceptance of LPI and modified LPI needs to be investigated with a consumer panel. SDS-PAGE and Bead-Assay indicated that the combination of enzymatic hydrolysis and fermentation of LPI is effective in breaking down large polypeptides into low molecular weight peptides and degrading with it the major allergen Lup an 1 of *L. angustifolius*. Thus, this two-step process represents a promising method for the reduction of the allergenic potential of LPI. Nevertheless, in vivo studies should be performed to investigate the allergenicity of fermented lupin protein hydrolysates. The use and possible applications of fermented lupin protein hydrolysates in food should be tested in practice by subsequent application trails.

Supplementary Materials: The following are available online at https://www.mdpi.com/2304-8158/10/2/281/s1, Table S1: Comparative retronasal aroma profile analyses of LPI and fermented LPI hydrolysates, Table S2: Intensity of bitter, salty and sour taste perception of LPI and fermented LPI hydrolysates, Table S3: Principal Component (PC) scaled scores and loadings for LPI and fermented LPI hydrolysates obtained by Principal Component Analysis (PCA).

Author Contributions: Conceptualization, K.S. and U.S.-W.; methodology, K.S., E.U. and U.S.-W.; formal analysis, K.S. and N.L.; investigation, K.S. and N.L.; resources, P.E.; writing—original draft preparation, K.S. and E.U.; writing—review and editing, K.S., N.L., E.U., U.S.-W. and P.E.; visualization, K.S. All authors have read and agreed to the published version of the manuscript.

Funding: This research received no external funding.

Institutional Review Board Statement: Not applicable.

Informed Consent Statement: Not applicable.

Acknowledgments: This work was supported by the Fraunhofer-Zukunftsstiftung.

Conflicts of Interest: The authors declare no conflict of interest.

References

1. *Word Population Prospects—The 2019 Revision*; Department of Economic and Social Affairs, Population Division: New York, NY, USA, 2020.
2. Eshel, G.; Shepon, A.; Makov, T.; Milo, R. Land, irrigation water, greenhouse gas, and reactive nitrogen burdens of meat, eggs, and dairy production in the United States. *Proc. Natl. Acad. Sci. USA* **2014**, *111*, 11996–12001. [CrossRef] [PubMed]
3. Bähr, M.; Fechner, A.; Hasenkopf, K.; Mittermaier, S.; Jahreis, G. Chemical composition of dehulled seeds of selected lupin cultivars in comparison to pea and soya bean. *LWT—Food Sci. Technol.* **2014**, *59*, 587–590. [CrossRef]
4. Bader, S.; Oviedo, J.P.; Pickardt, C.; Eisner, P. Influence of different organic solvents on the functional and sensory properties of lupin (*Lupinus angustifolius* L.) proteins. *LWT—Food Sci. Technol.* **2011**, *44*, 1396–1404. [CrossRef]
5. Purschke, B.; Meinlschmidt, P.; Horn, C.; Rieder, O.; Jäger, H. Improvement of techno-functional properties of edible insect protein from migratory locust by enzymatic hydrolysis. *Eur. Food Res. Technol.* **2018**, *244*, 999–1013. [CrossRef]
6. Meinlschmidt, P.; Sussmann, D.; Schweiggert-Weisz, U.; Eisner, P. Enzymatic treatment of soy protein isolates: Effects on the potential allergenicity, technofunctionality, and sensory properties. *Food Sci. Nutr.* **2016**, *4*, 11–23. [CrossRef] [PubMed]
7. Schlegel, K.; Sontheimer, K.; Hickisch, A.; Wani, A.A.; Eisner, P.; Schweiggert-Weisz, U. Enzymatic hydrolysis of lupin protein isolates—Changes in the molecular weight distribution, technofunctional characteristics and sensory attributes. *Food Sci. Nutr.* **2019**, *7*, 2747–2759. [CrossRef] [PubMed]
8. Akbari, N.; Mohammadzadeh Milani, J.; Biparva, P. Functional and conformational properties of proteolytic enzyme-modified potato protein isolate. *J. Sci. Food Agric.* **2020**, *100*, 1320–1327. [CrossRef]
9. Klost, M.; Drusch, S. Functionalisation of pea protein by tryptic hydrolysis—Characterisation of interfacial and functional properties. *Food Hydrocoll.* **2019**, *86*, 134–140. [CrossRef]
10. Schlegel, K.; Sontheimer, K.; Eisner, P.; Schweiggert-Weisz, U. Effect of enzyme-assisted hydrolysis on protein pattern, techno-functional, and sensory properties of lupin protein isolates using enzyme combinations. *Food Sci. Nutri* **2019**. [CrossRef]
11. Meng, S.; Tan, Y.; Chang, S.; Li, J.; Maleki, S.; Puppala, N. Peanut allergen reduction and functional property improvement by means of enzymatic hydrolysis and transglutaminase crosslinking. *Food Chem.* **2020**, *302*, 125186. [CrossRef]
12. Lqari, H.; Pedroche, J.; Girón-Calle, J.; Vioque, J.; Millán, F. Production of *Lupinus angustifolius* protein hydrolysates with improved functional properties. *Grasas y Aceites* **2005**, *56*, 135–140. [CrossRef]
13. Schindler, S.; Wittig, M.; Zelena, K.; Krings, U.; Bez, J.; Eisner, P.; Berger, R.G. Lactic fermentation to improve the aroma of protein extracts of sweet lupin (Lupinus angustifolius). *Food Chem.* **2011**, *128*, 330–337. [CrossRef] [PubMed]
14. Schindler, S.; Zelena, K.; Krings, U.; Bez, J.; Eisner, P.; Berger, R.G. Improvement of the Aroma of Pea (Pisum sativum) Protein Extracts by Lactic Acid Fermentation. *Food Biotechnol.* **2012**, *26*, 58–74. [CrossRef]
15. Schlegel, K.; Leidigkeit, A.; Eisner, P.; Schweiggert-Weisz, U. Technofunctional and Sensory Properties of Fermented Lupin Protein Isolates. *Foods* **2019**, *8*, 678. [CrossRef]
16. Mosse, J.; Huet, J.-C.; Baudet, J. Relationships between nitrogen, amino acids and storage proteins in *Lupinus albus* seeds. *Phytochemistry* **1987**, *26*, 2453–2458. [CrossRef]
17. Laemmli, U.K. Cleavage of structural proteins during the assembly of the head of bacteriophage T4. *Nature* **1970**, *227*, 680–685. [CrossRef]
18. Morr, C.V.; German, B.; Kinsella, J.E.; Regenstein, J.M.; Vanburen, J.P.; Kilara, A.; Lewis, B.A.; Mangino, M.E. A Collaborative Study to Develop a Standardized Food Protein Solubility Procedure. *J. Food Sci.* **1985**, *50*, 1715–1718. [CrossRef]
19. Lowry, O.H.; Rosebrough, N.J.; Farr, A.L.; Randall, R.J. Protein measurement with the Folin phenol reagent. *J. Biol. Chem.* **1951**, *193*, 265–275. [CrossRef]
20. Phillips, L.G.; Haque, Z.; Kinsella, J.E. A Method for the Measurement of Foam Formation and Stability. *J. Food Sci.* **1987**, *52*, 1074–1077. [CrossRef]
21. Wang, C.Y.; Johnson, L.A. Functional properties of hydrothermally cooked soy protein products. *J. Am. Oil Chem. Soc.* **2001**, *78*, 189–195. [CrossRef]
22. Goggin, D.E.; Mir, G.; Smith, W.B.; Stuckey, M.; Smith, P.A.C. Proteomic analysis of lupin seed proteins to identify conglutin beta as an allergen, Lup an 1. *J. Agric. Food Chem.* **2008**, *56*, 6370–6377. [CrossRef] [PubMed]
23. Sormus de Castro Pinto, S.E.; Neves, V.A.; Machado de Medeiros, B.M. Enzymatic Hydrolysis of Sweet Lupin, Chickpea, and Lentil 11S Globulins Decreases their Antigenic Activity. *J. Agric. Food Chem.* **2009**, *57*, 1070–1075. [CrossRef] [PubMed]
24. Huang, T.; Bu, G.; Chen, F. The influence of composite enzymatic hydrolysis on the antigenicity of β-conglycinin in soy protein hydrolysates. *J. Food Biochem.* **2018**, *42*, e12544. [CrossRef]
25. Kasera, R.; Singh, A.B.; Lavasa, S.; Prasad, K.N.; Arora, N. Enzymatic hydrolysis: A method in alleviating legume allergenicity. *Food Chem. Toxicol.* **2015**, *76*, 54–60. [CrossRef]
26. Meinlschmidt, P.; Schweiggert-Weisz, U.; Eisner, P. Soy protein hydrolysates fermentation: Effect of debittering and degradation of major soy allergens. *LWT—Food Sci. Technol.* **2016**, *71*, 202–212. [CrossRef]

27. Garcia-Arteaga, V.; Apestegui, M.; Muranyi, I.; Eisner, P.; Schweiggert-Weisz, U. Effect of enzymatic hydrolysis on molecular weight distribution, techno-functional properties and sensory perception of pea protein isolates. *Innov. Food Sci. Emerg. Technol.* **2020**, *65*, 102449. [CrossRef]
28. Muranyi, I.S.; Otto, C.; Pickardt, C.; Osen, R.; Koehler, P.; Schweiggert-Weisz, U. Influence of the Isolation Method on the Technofunctional Properties of Protein Isolates from Lupinus angustifolius L. *J. Food Sci.* **2016**, *81*, C2656–C2663. [CrossRef]
29. Vogelsang-O'Dwyer, M.; Bez, J.; Petersen, I.; Joehnke, M.; Detzel, A.; Busch, M.; Krueger, M.; Ispiryan, L.; O'Mahony, J.; Arendt, E.; et al. Techno-Functional, Nutritional and Environmental Performance of Protein Isolates from Blue Lupin and White Lupin. *Foods* **2020**, *9*, 230. [CrossRef]
30. Klupsaite, D.; Juodeikiene, G.; Zadeike, D.; Bartkiene, E.; Maknickiene, Z.; Liutkute, G. The influence of lactic acid fermentation on functional properties of narrow-leaved lupine protein as functional additive for higher value wheat bread. *LWT—Food Sci. Technol.* **2017**, *75*, 180–186. [CrossRef]
31. Tsumura, K.; Saito, T.; Tsuge, K.; Ashida, H.; Kugimiya, W.; Inouye, K. Functional properties of soy protein hydrolysates obtained by selective proteolysis. *LWT—Food Sci. Technol.* **2005**, *38*, 255–261. [CrossRef]
32. El-Adawy, T.A.; Rahma, E.H.; El-Bedawey, A.A.; Gafar, A.F. Nutritional potential and functional properties of sweet and bitter lupin seed protein isolates. *Food Chem.* **2001**, *74*, 455–462. [CrossRef]

MDPI

Review

Barley Protein Properties, Extraction and Applications, with a Focus on Brewers' Spent Grain Protein

Alice Jaeger [1], Emanuele Zannini [1], Aylin W. Sahin [1] and Elke K. Arendt [1,2,*]

[1] School of Food and Nutritional Science, University College Cork, T12 K8AF Cork, Ireland; 116458826@umail.ucc.ie (A.J.); e.zannini@ucc.ie (E.Z.); aylin.sahin@ucc.ie (A.W.S.)
[2] APC Microbiome Institute, University College Cork, T12 K8AF Cork, Ireland
* Correspondence: e.arendt@ucc.ie; Tel.: +353-021-490-2064

Abstract: Barley is the most commonly used grain in the brewing industry for the production of beer-type beverages. This review will explore the extraction and application of proteins from barley, particularly those from brewers' spent grain, as well as describing the variety of proteins present. As brewers' spent grain is the most voluminous by-product of the brewing industry, the valorisation and utilisation of spent grain protein is of great interest in terms of sustainability, although at present, BSG is mainly sold cheaply for use in animal feed formulations. There is an ongoing global effort to minimise processing waste and increase up-cycling of processing side-streams. However, sustainability in the brewing industry is complex, with an innate need for a large volume of resources such as water and energy. In addition to this, large volumes of a by-product are produced at nearly every step of the process. The extraction and characterisation of proteins from BSG is of great interest due to the high protein quality and the potential for a wide variety of applications, including foods for human consumption such as bread, biscuits and snack-type products.

Keywords: brewers' spent grain; barley protein; by-product valorisation; brewing waste; food ingredient

Citation: Jaeger, A.; Zannini, E.; Sahin, A.W.; Arendt, E.K. Barley Protein Properties, Extraction and Applications, with a Focus on Brewers' Spent Grain Protein. *Foods* **2021**, *10*, 1389. https://doi.org/10.3390/foods10061389

Academic Editor: Rotimi Aluko

Received: 30 April 2021
Accepted: 11 June 2021
Published: 16 June 2021

1. Introduction

Barley is the most commonly used grain in the brewing industry for the production of beer-type beverages. Both raw barley and malted barley are used, often in combination with adjunct grains such as rice and corn. Arguably, the most important fraction of the barley grain is the endosperm, comprised of starch granules suspended in a protein matrix, as it provides the growing plant embryo with all the products needed to begin growth. In this review, the proteins of the barley grain will be the focus. In particular, the extraction and application of barley proteins from brewers' spent grains will be explored. As brewers' spent grain is the most voluminous by-product of the brewing industry, the valorisation and utilisation of spent grain protein is of great interest in terms of sustainability. There is an ongoing global effort to minimise processing waste and increase up-cycling of processing side-streams in order to support sustainable growth in the coming decades. However, sustainability in the brewing industry is complex. There is an innate need for a large volume of resources such as water and energy, as well as a disconnect between raw materials and the processing facility resulting in an increased carbon footprint [1]. In addition to this, large volumes of by-product are produced at nearly every step of the process including spent grain, spent yeast, wastewater and spent hops, among others. The brewing industry is responsible for a huge volume of waste with ~10,000 tonnes of liquid waste and 137–173 tonnes of solid waste produced per 1000 tonnes of beer [2], so the potential valorisation of these waste products is of great economic interest. Due to its high nutritional value, particularly in terms of high-quality protein, the application of BSG proteins in human nutrition is of particular interest. The addition of BSG has already been examined in breads, biscuits and other bakery products as well snack-type products. In addition to this, applications of BSG include use in animal feed formulation for a range of species [1–4] as well as use as a potential substrate for other industrial processes, namely biogas production, polysaccharide extraction and phenolic compound isolation [4–6].

2. Barley Structure

Barley (*Hordeum vulgare, vulgare* L.) is a widely cultivated and easily adaptable crop, most commonly used as a raw material for the brewing process. The main components of the barley grain are the embryo, the aleurone layer, the endosperm and the husk (grain covering) (Figure 1). The embryo is the most important living tissue in the grain and develops into the plant. It contains a limited amount of starch, lipids and protein to sustain the embryo prior to germination [7]. The aleurone layer is two to three cells thick and encases the endosperm. It is a living tissue containing protein, phytin, phospholipids, RNA and some carbohydrate and plays a critical role in the regulation of endosperm degrading enzymes [7–9]. The grain covering acts to protect the integrity of the grain throughout harvesting and processing. The grain covering can further be divided into the seed coat, pericarp and husk [10]. The husk consists of mainly cellulose and a small amount of polyphenols and bitter substances, whereas the pericarp develops from the ovary wall and acts as a protective cover over the kernel [9]. The husk has proven to aid the filtration process during the lautering step of brewing as well as protecting the grain during malting [7]. However, husk-less cultivars have been shown to be effective in alcohol production, although with an altered malting process [11].

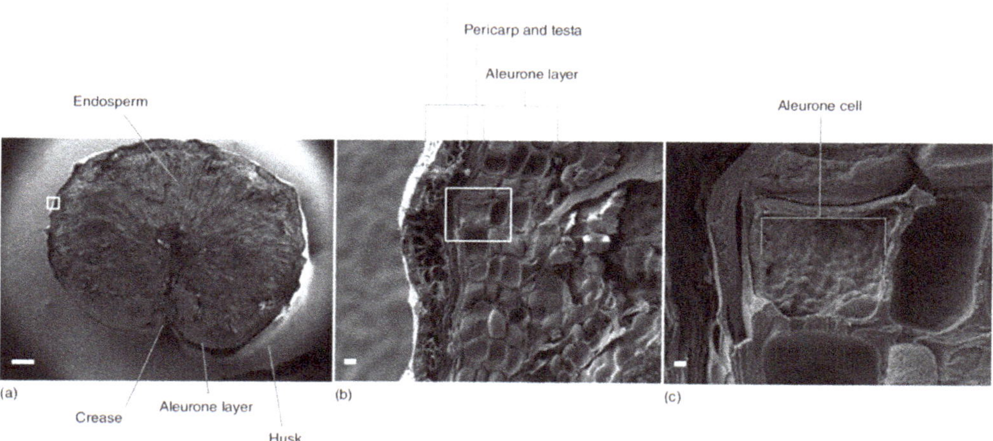

Figure 1. Structure of the barley kernel sectioned transversely (**a**) with detailed husk (**b**) and starchy endosperm (**c**), Reprinted with permission from ref. [9]. Copyright 2013 Elsevier.

The endosperm consists of starch granules suspended in a protein matrix. The starch is the most abundant component, comprising approximately 60% of grain weight. The endosperm cell walls consist of mainly non-starch polysaccharides, namely β-glucans and arabinoxylans [7].

The barley grain is reported to contain between 8 and 30% protein as a percentage of total mass [7,12–14]. This protein is synthesized in the endosperm and the aleurone layer during grain development and accumulates during grain filling. The crude protein content is used to predict malting and, therefore, brewing quality [12,15]. The protein content of the barley grain has a complex relationship with quality with regards to malting barley. While high protein barley would be desirable for feed applications, a lower level is desirable for malting varieties. The ideal protein content for malting barley resides between 10 and 12%, and too high or low a protein content can negatively affect malting quality and, therefore, brewing capacity. In general, a high protein content is undesirable, as there is a strong correlation between a high protein content and a low carbohydrate content, leading to a low extract yield. This may also lead to excessive enzyme activity. However, too low a protein level may lead to insufficient amino acids for the yeast nutrition during the brewing process as well as low enzymatic activity, leading to a lower level of fermentable sugars and, therefore, a poor extract yield [7,15–17].

The protein content of the barley grain is highly variable and can be affected by barley cultivar as well as environmental conditions and the addition of fertilizers. In hot, dry and nitrogen-rich environments, it has been shown that the protein content of the barley grain is increased, bringing the protein level above that which is suitable for malting (approx. 11.5% protein) [16–19]. Increased temperature and reduced rainfall have been shown to have a negative correlation with malting quality as measured by malt extract [15]. However, very high levels of precipitation are also unfavourable, leading to reduced endosperm development [18]. This is undesirable for brewers, as an increased amount of malt will need to be used to produce the beer. Therefore, the development of new barley cultivars to produce grain with a high malt extract and diastatic power while keeping an acceptable protein level is of great interest. Diastatic power refers to the combined activity of α- and β-amylases and β-glucanases. These enzymes play a crucial role in the brewing process, degrading malt carbohydrates to fermentable sugars [17,20].

The application of a nitrogen (N) fertilizer also has a significant effect on barley protein levels. Nitrogen addition was observed to be raising grain protein concentration and reducing malt extract, with a much more significant effect in warmer and drier regions [16]. However, the use of low protein barley genotypes may allow for barley with an acceptable grain protein level to be grown on land with increased nitrates (NO_3). It has been shown that, while a low-protein genotype is generally associated with reduced diastatic power, the addition of nitrogen fertilizer can increase this level to be similar to that of the most commonly grown malting barley cultivars [17]. The genotype, growing climate, soil composition and fertilizer addition can be of great importance, influencing the performance and quality of the barley malt during the brewing process.

Genetic modification has also been considered for the development of the ideal barley cultivar. While this method does show potential, it cannot yet compete with the more broadly used cultivars in terms of yield [21].

3. Barley Proteins

3.1. Hordeins/Prolamins

The majority (30–50%) of the barley protein fraction consists of hordeins, a protein belonging to the prolamin group, so named due to their high content of glutamine and proline [14,15,22]. These hordeins (or prolamins) are not single proteins but complex polymorphic mixtures of polypeptides [23]. The other barley proteins consist of a mix of albumins, globulins and glutelin [7]. The hordeins can be further divided into subsections based on their amino acid compositions and extractability characteristics [12,24]. B-hordein is the largest fraction accounting for between 70 and 90% of the total hordein content, and it is the major storage protein of the barley grain. B-hordein can be further broken down into subunits B1, B2 and B3 based on their electrophoretic capabilities. Other fractions include C-, D- and γ-hordeins, with C-hordeins accounting for 10–20%, and the remaining types making up less than 5% collectively [12]. B-hordeins and γ-hordeins are sulphur-rich, while C-hordeins are sulphur-poor. D-hordeins are distinguished by their high molecular weight (>100 kDa) [24]. These differing proteins are coded by specific genes: Hor2 (B-hordeins), Hor1 (C-hordeins), Hor3 (D-hordeins) and Hor5 (γ-hordeins [24]. A study by Howard et al., 1996, observed that the measurement of D-hordein levels in the barley grain gave a more significant single measurement for malting quality spanning a range of climatic and agronomic conditions, compared to total grain protein [15].

3.2. Glutelin

Glutelin is the second most abundant fraction in barley storage proteins, making up 35–45% of the total storage protein. It contains high levels of glutamine, proline and glycine, while also being rich in other hydrophobic amino acids [25]. In a study by Wang et al. (2010), barley glutelin was found to have very good emulsifying properties [26]. This is theorised to be due to a balanced ratio of polar and non-polar amino acids, allowing the protein to adsorb easily onto the surface of oil droplets and quickly lower the interfacial tension. However, due to poor solubility, glutelin requires dehydration at pH 11 and further pH adjustment for the emulsification properties to be active, which is not practical for applications in food systems. Deamidation has been found to increase barley glutelin solubility and emulsifying properties [25].

3.3. Protein Z

Protein Z is the major barley albumin with a molecular mass of 40 kDa and a pI ranging from 5.5 to 5.8 [27]. Protein Z is a member of the serpin protein group and represents about 5% of total barley albumin. However, protein Z is heat stable and resistant to enzymatic degradation, meaning it survives the brewing process unmodified and is one of the major proteins still present in the finished beer, with a particular role regarding beer foam stability alongside lipid transfer proteins [27,28]. The relationship between the concentration of protein Z and foaming properties has been studied in a few separate instances. It was found that while protein Z is a family of proteins, protein Z4 was the predominant type, as it made up 80% of the protein Z fraction [28]. It was also determined that protein Z plays an integral role in foam stability, as the addition of purified protein Z enhanced foam stability. Similarly, Ayashi et al. (2008) found that protein Z4 and lipid transfer protein 1 (LTP-1) were beneficial to foaming capacity and stability. It is theorised that LTP-1 does this by binding lipids, which are known to negatively influence foam stability [29], while protein Z4 interacts with the α-acids derived from the hops addition. While both the albumin and hordein fractions of the barley protein are capable of foaming, the albumin fraction has significantly higher foam stability. However, denaturation of these protein fractions causes an increase in hydrophobicity, therefore enhancing foaming properties, with albumins being more stable than hordeins. The foam stability of both fractions is improved by exposure to the bitter α-acids derived from the hops, but both protein fractions are also susceptible to hydrolysis by yeast-derived proteinase A, resulting in diminished foam stability [30].

A study by Boba (2010) examined the use of protein Z as a method for monitoring the malting progress of barley grain via glycation. The proteins in barley malt are glycated by D-glucose, a product of starch degradation during malting, and potentially improve foam stability and prevent protein precipitation during wort boiling [31]. It is the lysine and arginine residues that are glycated, with 16% of the lysine residues in protein Z being modified during the Maillard reactions of the malting process. Prolonging of malting increases the level of glycated peptides. Therefore, measuring its concentration using gel electrophoresis, liquid chromatography and mass spectrometry can be used as an efficient measurement of malting progression [31]. This could be applied effectively to control protein variety and quantity in the final beer.

3.4. Effect of Brewing Processes on Barley Proteins

The brewing process significantly alters the proteins of the grains. An understanding of the changes to barley proteins that occur during malting and mashing is key to determining how these processes affect the final protein composition of BSG as well as the residual beer proteins.

3.4.1. Malting

Malting is the controlled germination of the barley grain. During malting, partial degradation of the cell walls, starchy endosperm and storage protein occurs [32,33]. The hydrolytic enzymes responsible for this process are produced by the grain during the first stages of germination. For protein degradation, endoproteases and carboxypeptidases are produced and secreted into the endosperm. The endoproteases act to break down the proteins into peptides, while the carboxypeptidases further degrade the peptides into free amino acids [33]. Greater than 70% of the proteins, mainly hordeins and glutelin, in the barley grain are degraded by endoproteases to create smaller peptides during the malting process [33]. A study by Celus et al. (2006) demonstrated that B- and D-hordeins are the main fractions degraded [32]. Adequate modification of the grain constituents during malting is key to the final quality of the beer. Insufficient modification can cause issues such as low extract yields and low fermentability [33].

3.4.2. Starch Degradation Inhibition

Insufficient protein degradation during malting has been shown to inhibit the degradation of starch during the mashing process. A study by Slack et al. (1979) examined this phenomenon and determined that hordein proteins are very closely associated with the surface of the starch granules and can form a protective layer around the granule. Due to this, starch degrading enzymes (namely α-amylase) have limited access to the starch granules, therefore limiting the production of fermentable sugars. It was shown that this effect is even more pronounced for the small starch granules in comparison to the large starch granules. During mashing at approx. 65 °C, the larger starch granules gelatinise, therefore becoming more available to the α-amylase to be converted to maltose. Small starch granules have a higher gelatinisation temperature, and so, the same effect is not observed [34]. During malting, the B- and C-hordeins on the starch granule surface should be largely degraded, allowing for the starch degrading enzymes to easily access the granule. However, in cases where the malt is under-modified or raw adjuncts are introduced, the inhibitory effects of hordeins can be observed. A study by Yu et al. (2018) investigated the same effect and confirmed that barley proteins, in particular the soluble component, play a role in slowing starch degradation. This study also looked at alternate methods by which this occurs [35]. As well as the physical barrier as described by Slack et al. (1979), it is also theorised that the digestive enzyme (α-amylase) binds to the hordein and glutelin and is, therefore, unable to act on the substrate. It is thought that proteins adhere to the starch granules via disulphide linkages, as proteins can be removed by the addition of cysteine [34].

3.4.3. Mashing

Mashing is the first step in the brewing process and continues the enzymatic degradation that began during the malting of the barley grains. Most endoproteases (approx. 90%) will survive the kilning process and will, therefore, be present in the mashing phase [33]. Therefore, protein degradation into peptides and amino acids will continue during mashing. A study by Jones and Marinac (2002) investigated the effect of mashing on malt proteinase activity. It found that the endoproteases maintained a high activity level during the protein rest stage of mashing, where the mash is held at approx. 38 °C to aid protein degradation. However, protease activity rapidly deteriorates when the temperature is raised to 70 °C. This rise in temperature is to facilitate the hydrolysis of malt and adjunct starches to sugars, meanwhile deactivating the proteases. Due to this, soluble protein levels in the wort can be manipulated by the brewer, if so desired, by increasing or decreasing the length of the protein rest phase. It is also noted that attempting to change the ratio of protein fractions within the soluble protein is not likely to succeed, as all of the proteinases appear to denature at the same temperature [36]. Due to the near-total deactivation of the proteases during the starch conversion phase, elongation/shortening of this step would have no effect on wort protein levels. During mashing, a significant fraction of grain proteins are degraded and solubilised, and through wort boiling, they are glycated and coagulated to form aggregates that can be separated during wort boiling as 'hot trub' [37]. A study by Celus et al. (2006) determined that during mashing, disulphide bonding is induced between B- and C-hordeins. While it was previously thought that B- and D-hordein (and glutelin) were the main components of the aggregate, this study suggests that it is instead composed of mainly B-hordeins, in which some C-hordein is enclosed [32]. These aggregated hordeins have been reported to cause mash separation issues [32,34,35,38]. An earlier study by Moonen et al. (1987) also described the formation of this protein complex during mashing, consisting of residual malt proteins and glutelins. It also examines how this complex negatively impacts the lautering and filtration processes [38]. This once again highlights the importance of sufficient protein degradation during the malting stage, as poor-quality malt with insufficient modification will contain a higher level of residual protein as well as a lower enzyme activity, therefore causing more issues regarding gelation.

This impenetrable protein complex persists as a coating on the residual spent grains, the upper section of the 'oberteig' layer. Therefore, they are an important fraction of the brewers' spent grain (BSG) protein complement. Due to the impenetrable nature of the matrix, protein extraction may need to involve pre-treatments and/or vigorous disruption procedures [39].

The methods behind the formation of these aggregates are largely unknown. Moonen et al. (1987) hypothesised that high-molecular-weight subunits (such as D-hordeins) form the 'backbone' structure of these gels. This idea is supported by Skerritt and Janes (1992), who used protein assays, electrophoresis and HPLC to examine the relationship between these disulphide-linked protein aggregates and malting quality. It was determined that D-hordeins were the slowest to extract and are, therefore, assumed to act as the 'backbone' of the gel matrix [40]. An increased level of sodium dodecyl sulphate (SDS) unextractable proteins, namely B1- and B2-hordeins, indicates a lower-quality malt, as hordein subunits are more readily and easily extracted from good-quality malt [40].

3.4.4. Residual Beer Proteins

Many studies have focused on comparing the protein profiles of malts with the protein profiles of the final beer product. One such study by Klose et al. (2010) examined the protein profiles of the malt and beer as well as the intermediate stages (wort and hot trub) using two-dimensional gel electrophoresis as well as lab-on-a-chip technology. It was determined that the 500 mg of proteinaceous material present [41] in the final beer originated from the initial grain, where increased inherent levels of cysteine and/or lysine allowed the proteins to be resistant to enzymatic degradation and increased their heat stability [27]. This need for heat stability was also noted by Curioni et al. (1995), where the major fraction of residual protein in beer was determined to comprise of two polypeptides with an approximate molecular weight of 40 kDa, later determined to be two albumins coinciding with the characteristics of protein Z [41]. It was also noted that upon examination of the precipitated protein in the 'hot-trub', the molecular mass of the peptides was low, indicating that the enzymatic degradation of proteins established during malting continued throughout mashing [27].

4. Overview of BSG

Brewers' spent grain (BSG) is the most abundant by-product of the brewing process, consisting of up to 85% of total brewery waste [6]. On average, 20 kg of wet BSG is produced for every 100 L of finished beer [10]. Due to an ever-increasing interest in waste reduction and by-product valorisation, the fractionation and application of BSG-derived ingredients is an expanding field of study. BSG is what remains of the initial brewing grains after being subjected to the malting and mashing processes. Therefore, it consists of mainly the husk–pericarp–seedcoat, meaning BSG is rich in both cellulosic and non-cellulosic polysaccharides, as well as lignin. While rich in fibre, it is also rich in protein, with the protein fraction accounting for approx. 19–30% of total grain composition as outlined in Figure 2 [32,42–49].

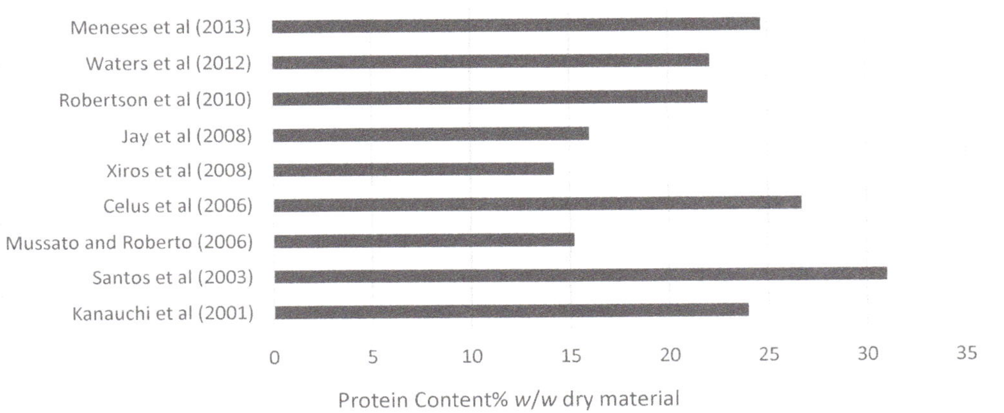

Figure 2. Protein content of brewers' spent grain (BSG), % *w*/*w* dry matter [32,42–49].

As well as a high protein content, BSG also contains a high level of essential amino acids, as outlined in Table 1. These essential amino acids constitute approx. 30% of the total protein, and lysine is of particular interest, as it is generally the limiting amino acid in cereal foods for human consumption [48]. In a study by Connolly et al. (2013), protein isolates prepared from pale and black BSG were characterised for amino acid composition. Glutamine and proline were the most abundant amino acids present in both isolates, and different temperatures were used during the protein extraction to determine the effect of temperature. The results showed that the pale BSG, extracted at 50 °C, contained the highest levels of all amino acids, except for cysteine [39]. Glutamine, proline and leucine were the most abundant, while the sulphur-containing amino acids, methionine and cysteine, were the scarcest. The isolate extracted at a lower temperature (20 °C) had a lower amino acid content overall. However, it was similar to the 50 °C in that glutamine and proline were the most abundant, while the sulphur-containing amino acids were present in the lowest levels. The black BSG isolates both contained a lower level of amino acids than their pale counterparts. This is most likely due to the roasting of the malt at high temperatures (>200 °C), where amino acids may be degraded or used in Maillard-type reactions. Essential amino acids are crucial for human health and can be lacking in certain foods, therefore requiring fortification. Due to the presence of a high level of essential amino acids, BSG-derived proteins as fortification agents in foodstuffs for human consumption present an economical solution to these issues.

Table 1. Amino acid composition of barley, barley malt and BSG—adapted from Waters et al. (2012).

	Barley	Malt	BSG
Total protein (% w/w)	9.65	8.52	22.13
(% of total)			
Aspartic acid	0.19	0.17	4.81
Glutamic acid	0.85	0.75	16.59
Asparagine	0.23	0.33	1.47
Serine	0.12	0.07	3.77
Glutamine	ND	ND	0.07
Histidine	1.59	1.9	26.27
Glycine	0.08	0.06	1.74
Arginine	0.21	0.23	4.51
Alanine	0.22	0.23	4.12
γ- aminobutyric acid	2.56	0.01	0.26
Tyrosine	0.14	0.14	2.57
Valine	2.56	0.24	4.61
Threonine	0.01	0.02	0.71
Methionine	0.03	ND	ND
Tryptophan	0.01	ND	0.14
Phenylalanine	0.2	0.21	4.64
Isoleucine	0.17	0.17	3.31
Leucine	0.3	0.29	6.12
Lysine	2.52	3.69	14.31

4.1. Adjuncts

Adjuncts are ingredients, other than malted barley, used in brewing to provide additional carbohydrates to contribute to sugars in the wort. While the proteins of the barley grain are the focus of this review, if discussing proteins in brewers' spent grains, it is important to consider that alternative protein sources may also be present.

In certain countries, namely Germany, Switzerland and Greece, the purity law ('Reinheinsgebot') states that beer may only be made from water, malt, hops and yeast, so adjuncts are prohibited. Generally, adjuncts are used to reduce raw material cost and/or to add certain desirable qualities to the finished beer [50,51]. While they are predominantly cereal-based, sugar-based syrups can also be used. The most common cereals used as adjuncts are rice and maize, as well as raw, un-malted barley. Due to differing gelatinization temperatures, adjuncts often have to be 'cooked' separately from the main mash in a double mash conversion system [52]. However, triticale can be added into the main mash as it has a lower gelatinisation temperature of 59–65 °C, well within normal mashing temperatures [53]. A study by Agu (2002) also investigated the possible use of sorghum as a brewing adjunct alongside barley. It was discovered that sorghum has the potential to release higher levels of peptides than commonly used maize and was very effective at a 5% addition, whereas at 20% addition, a decrease in total wort peptides was noted [54]. A separate study by Glatthar et al. (2005) investigated the use of un-malted triticale as a brewing adjunct and noted that the wheat and rye hybrid held great potential for use as a brewing adjunct. While adjunct addition levels vary widely, high levels of addition can present increased effects on sensory characteristics as well as technological difficulties, such as increased viscosity [51,53,55]. In a study by Yorke et al. (2021), it was found that while 30% adjunct addition had little to no observable effect on beer characteristics, 60% addition presented increased sensory differences, low free amino nitrogen and dramatically altered the fermentation profile [51].

The protein contents of adjuncts can greatly affect wort quality. The balance between adjuncts and malt must be carefully monitored as adjunct addition can 'dilute' the enzymatic activity of the malt, therefore requiring a malt of a higher diastatic power or the addition of commercial enzymes [52]. A study carried out by Schnitzenbaumer et al. (2012) investigated the effect of using oats as a replacement for malted barley. It was noted that the replacement of 20% or more of malted barley with oats resulted in a decrease in the amount of free amino acids and total soluble nitrogen, as well as decreased extract and significantly reduced foam stability.

Due to the frequent use of adjuncts in the brewing industry, the analysis, extraction and utilization of brewers' spent grain protein is not only dealing with barley proteins but also potentially those of adjuncts used in the brewing process.

4.2. Rice Protein

Rice is a commonly used adjunct in brewing to produce a light and clean tasting beer. Due to its low cost and high starch content, it is used to supplement the carbohydrates available from the barley and malt, leading to increased fermentative capability. Brewers' rice is generally produced as a by-product of the edible rice milling industry, as up to 30% of the grains may be fractured and deemed unsuitable for the edible rice market [56]. While the starchy endosperm is the main component of the rice grain (89–94%), brown rice contains 6.6–7.3% protein, milled rice 6.2–6.9% protein, and basmati rice contains 8.2–8.4% protein [57]. Although levels can vary greatly depending on the environment, soil and cultivar, in general, this protein level is lower than that of both barley, barley malt, maize and sorghum [58]. As opposed to the barley grain, where the major proteins are the endosperm-specific prolamin storage proteins, the major protein in the rice grain is a glutelin-type storage protein, containing 63.8–73.4% glutelin. Other proteins present include water-soluble albumins (9.7–14.2%), salt soluble globulins (13.5–18.9%) and alcohol soluble prolamins (3.0–5.4%) [57]. Rice proteins have been shown to be significantly resistant to hydrolysis [59]. As a result, the addition of a rice adjunct provides very little free amino nitrogen (FAN), and this must be compensated for by the barley and malt fractions.

4.3. Maize Protein

Alongside rice, maize is another grain extensively used as an adjunct in the production of beers. In a similar way to rice, the high volume of starchy endosperm serves to supplement the sugars provided by the barley and barley malt alone for a more efficient and economical production process. A study by Agu (2002) describes the effects of maize addition at 5–20%. The results showed a decrease in extract recovery but an increase in levels of FAN and peptide nitrogen when compared to barley adjuncts [54]. Maize contains ~10% protein [60]. Similarly to barley, the major proteins present in maize are endosperm-specific prolamin storage proteins. These are mainly small, 19–25 kDa α-prolamins [58].

5. Protein Extraction Methods

5.1. Innate Protein Extraction Methods

Several methods of protein extraction from the innate barley grain have been explored in the literature to date. Generally, grains are milled and/or pearled before being exposed to an extraction buffer [26,61,62]. Extraction solutions for barley protein fractions include salt, alcohol and alkaline. In a study by Wang et al. (2010), the hordein fraction was isolated by using an ethanol solution (55–75% v/v), whereas the glutelin fraction was extracted from the residues by application of an alkaline solution (pH 9–11.5). The hordein fraction was isolated using a rotary evaporator to remove ethanol, while the salt and alkali solubilised fractions were adjusted to pH 5 to facilitate protein precipitation [26].

5.2. Extraction of Proteins from BSG

In order to utilise each of the valuable components of BSG, methods for separating the fractions must be determined. In order for barley proteins to be successfully commercialised and used in an industrial setting, efficient extraction methods are needed to separate the valuable proteins from the spent grain. Some of the potential protein extraction methods include alkaline extraction, acid extraction and filtration, as well as more novel techniques such as ultrasonic treatment and pulsed electric field treatment.

5.3. pH Shift Extraction Methods (Alkaline Extraction)

Alkaline extraction is the most widely used and well-known methods for protein extraction. A study by Cavonius et al. (2015) explains the mechanics of this method well. At an alkaline pH, the proteins in a system obtain a net negative charge, increasing repulsion within and between the protein molecules. The interactions between the protein molecules and water are promoted; therefore, the proteins are solubilised. When the pH is lowered to the proteins isoelectric point (pI), this negative charge is lost and the protein's interaction with the water is minimized. This leaves the proteins insoluble and allows them to precipitate out of solution [63]. In this study by Cavonius et al. (2015), the 'pH-shift' method was applied to microalgal biomass, but the same principle can be applied to BSG.

Alkaline extraction was utilised in a study by Celus et al. (2007) to create a BSG protein concentrate (BPC). A solution of 0.1 M NaOH was used, and the mixture was held at 60 °C for 60 min to allow for optimum extraction. Following this, the proteins were precipitated by adjusting the pH of the filtrate to pH 4.0 using 2.0 M citric acid, and this precipitate was separated using centrifugation. The resulting BPC contained 60% protein on a dry weight basis [64].

A more recent study by Connolly et al. (2013) explored and characterised the protein-rich isolates from wet pale and wet black brewers' spent grain using alkaline extraction. While KOH and Na_2CO_3 were considered for use, NaOH was determined to be the most effective. It was determined that 110 mM was the most effective and efficient concentration of NaOH for extraction, even though extracted protein peaked at 200 mM. Several other parameters for optimal protein extraction were also explored in this study. Generally, protein extraction for both types of BSG increased over a range from 20–60 °C with pale BSG increasing from 37.17 to 88.20 mg g^{-1} and black BSG increasing from 37.02 to 64.32 mg g^{-1} [39]. Temperatures over 60 °C were not considered due to the risk of protein denaturation. This is in agreement with previous studies by Bals et al. [65] and Celus et al. [64]. The optimum weight/volume ratio was also determined to be 1:20. The barley proteins present in the alkaline extracts of BSG consisted mainly of hordeins.

A study by Diptee et al. (1989) explored the various parameters involved in protein extraction yield from brewers' spent grain using response surface methodology. These variables included time, temperature and particle (grain) size. While the efficacy of commonly used protein extraction solutions on BSG proteins has been debated, the extraction solution used in this study consisted of 3% sodium dodecyl sulphate and 0.6% Na_2HPO_4, followed by precipitation of the protein in ethanol. Using this method, a 60% protein yield was obtained using a BSG: extractant ratio of 2.5:100 and by heating the mixture to 90 °C for 95 min [66]. This was in agreement with the maximum value predicted using response surface methodology.

As seen through extensive use, alkaline extraction is a viable and efficient method to extract proteins, namely hordeins and glutelin, from BSG. The protein obtained from BSG could potentially be used in a variety of applications, including the food and nutraceutical industries.

5.4. Filtration

Filtration, including microfiltration and ultrafiltration, has been a commonly employed method of separating out fractions based on molecular size. Examples include the use of membrane separation to separate the solid fractions of corn, resulting in high protein corn gluten meal (67% protein), high-fat corn germ, corn starch and high fructose corn syrups [67].

A study by Tang et al. (2009) applied ultrafiltration to brewers' spent grain as a method of protein extraction. A BSG extract was prepared using ultrasound-assisted extraction with a sodium carbonate buffer, then filtered through a nylon cloth and centrifuged. The resulting mixture was used as the feed solution for ultrafiltration. Protein recovery by ultrafiltration was highly successful, with more than 92% of the protein being retained using both 5 and 30 kDa membranes [68]. The protein contents of the final products were

20.09% protein when concentrated using the 5 kDa membrane and 15.98% protein when concentrated using the 30 kDa membrane [68]. Connolly et al. [69] also used membranes of 10 and 30 kDa to concentrate proteins from a BSG preparation. A benefit of using this type of fractionation is the lack of added heat, yielding a higher-quality protein product.

5.5. Pre-Treatments

The effects of various pre-treatments of the BSG have also been widely investigated.

Shearing as a pre-treatment proved to be beneficial. It is thought to be due to the physical disruption of cell membranes and lignocellulosic material that allows for easier extraction of proteins. The use of acid as a pre-treatment was also investigated, but in this study, it did not appear to have any effect on protein extractability [39]. Hydrogen peroxide proved to be effective, as its application increased protein yield by 27%. However, this method of pre-treatment is not frequently used due to its potential to cause oxidation of the extracted protein.

A study by Qin et al. (2018) outlines the effects of several differing pre-treatment strategies on the protein extractability of BSG. A combination treatment of alkaline and acid seemed to be effective, with up to 95% protein extraction achieved [70]. However, it was discovered that a high percentage of lignin and carbohydrates were solubilised with the protein, meaning the process was not selective for proteins only. This would limit its practical applications, as a purer protein fraction is usually the most desirable. A single step dilute acid pre-treatment using H_2SO_4 was proven to be effective at extracting up to 90% of the BSG protein; however, a similar issue with the carbohydrates and lignin was present. The benefits of hydrothermal pre-treatments were examined at differing solid to liquid ratios, differing temperatures (30 to 135 °C) and increasing time (1 and 24 h). It was found that this allowed for 64–66% protein extraction and was more selective for proteins than the acid and alkaline treatments [70]. Hydrothermal treatment is potentially a very good option for optimisation of protein extraction from BSG, as it requires no addition of chemicals and can be carried out at a relatively low temperature to preserve protein quality but also minimise energy use when compared to acid treatment.

5.6. Enzymatic Treatment

The use of enzymes, namely proteases, to separate proteins from BSG has also been explored in literature. In addition to alkaline treatment, enzymatic treatment could optimise protein extraction potential from BSG. A study by Niemi et al. (2013) states that while up to 53% of proteins in BSG can be solubilised in alkaline conditions without protease addition, up to 76% was solubilised with the addition of an alkaline protease, in addition to carbohydrase treatment.

Alcalase, an alkaline protease extracted from *B. licheniformis*, has been reported to be effective in aiding to separate the protein fraction from BSG [64,71–73]. Determined to be capable of solubilising and extracting up to 77% of BSG proteins, creating peptides with a much lower molecular weight than other less effective proteases [73]. It was found that Alcalase performed better at pH 8.0, requiring a pH adjustment of the raw materials. To achieve maximum protein solubilisation, 10–20 μL of enzyme per g dry matter was required to work for approx. 4 h. However, significant protein separation, up to 64%, was observed using does as low as 1.2 μL [73]. It was also noted through amino acid analysis that an Alcalase treatment was particularly efficient at solubilising the proline and glutamine components of barley hordeins, more so than other amino acids. A study by Qin et al. (2018) also aimed to utilise the protein solubilisation effects of Alcalase for protein extraction optimisation from BSG. However, the results differed from those reported in the literature, with only 43–50% protein extraction. It was theorised to be due to the pH of the mixture being slightly too low at pH 6.25 after pre-treatment as opposed to pH 6.5, the optimum pH for Alcalase. Due to this, results differed from those reported in the literature. When the treatment was repeated at pH 8.0, a minor increase in protein solubility was observed, though it was still a lower figure than had been reported previously.

A study by Niemi et al. (2013) investigated the combined use of proteases and carbohydrases to maximise protein solubility. Milled BSG was treated with Depol 740 L in a 10% weight/volume mixture for 5 h at 50 °C with constant stirring. This BSG was then treated with different proteases, including Alcalase, ProMod and Acid Protease A [71]. Alcalase was found to be the most effective enzyme at pH 9.5, solubilising 35% of BSG biomass, while the other proteases only solubilised approx. 14%. It was found that enzymatic carbohydrate digestion was beneficial in maximising proteins solubilisation and extraction from BSG, leading to an increased protein yield. It is theorised to be due to the degradation of important cell structures by the carbohydrate degrading enzymes, allowing greater access to the internal cell structures. It was also noted that prolonged exposure to alkaline conditions, in the absence of a protease, also led to a significant amount of protein being solubilised, indicating that prolonged exposure to a high pH decreases cell wall integrity and increases protein solubility [71]. Depol 740 was also used in combination with Alcalase to improve protein solubilisation in a study by Treimo et al. (2008). While it was found to improve the overall solubilisation of the biomass, the increase was largely in the solubilisation of non-proteinaceous material. Overall, treatment with carbohydrate degrading enzymes had only minor effects on protein solubilisation, as opposed to the larger effect observed by Niemi et al. (2013).

A study by Celus et al. (2007) investigated the technological properties of BSG protein hydrolysates as prepared by enzymatic hydrolysis. Commercially available enzymes, including Alcalase, Flavourzyme and Pepsin, were used over a range of times and concentrations to obtain hydrolysates with varying degrees of hydrolysis (DH). BSG protein concentrate was first prepared by alkaline extraction and then subjected to enzymatic hydrolysis. It was determined that enzymatic hydrolysis generally improved the solubility of BSG proteins at lower pH values, leading to increased protein yield across the entire pH range [64]. A significant difference in technological function was also observed between the samples, based on the enzyme used during the hydrolysis. Emulsification and foaming properties decreased with increasing DH when Alcalase or Pepsin was used, as these properties are reliant on peptides with a MW in excess of 14.5 kDa. However, the opposite was observed for hydrolysates prepared with Flavourzyme [64]. Hydrolysis of the BSG proteins greatly improved their technological functionality. Foaming and emulsification properties, which were negligible in the raw BSG, were significantly improved by the hydrolysis treatment.

The use of thermochemical pre-treatment, as well as carbohydrate and protein hydrolysing enzymes on the fractionation of protein and lignin from BSG was investigated in a study by Rommi et al. (2018). Significant protein solubilisation was achieved using an alkaline protease treatment, bringing BSG protein solubilisation from approx. 15% to almost 100%, in agreement with the studies mentioned previously [64,71–73]. The steam explosion treatment reduced protein solubility but increased the efficacy of extract separation in the centrifugation step. Lignin and protein from BSG co-extracted and could only be partially separated using acidic precipitation. Due to this, raw BSY may be preferred in the production of BSG protein hydrolysates, with an aim to limit lignin co-extraction [72].

5.7. Ultrasonic-Assisted Extraction

Ultrasonic treatment is a more novel method that has been explored as a means of aiding protein extraction from BSG. In a study by Tang et al. (2010), the optimum ultrasound treatment was determined using response surface methodology, regarding three parameters: time, power and solid/liquid ratio. A treatment time of 81.4 min, at an ultrasonic power of 88.2 W/100 mL extractant and a solid/liquid ratio of 2.0 g/100 mL, was used to predict an optimal yield of 104.2 mg/g BSG, and this was in accordance with the experimental value [74]. Ultrasonic treatment of BSY was also determined to improve certain functional characteristics, namely fat absorption capacity, foaming properties and emulsifying properties, as well as improving extraction yield [75]. When ultrasound treatment is used in combination with enzymatic treatment, as in Yu et al. (2019), enzymatic loading and incubation time can be reduced, and protein solubility significantly increases [76].

5.8. Pulsed Electric Field

Pulsed electric field technology (PEF) is another novel extraction method for food compounds that is gaining interest. PEF consists of the application of short-duration pulses of an electric current through a sample that is secured between two electrodes [77]. The enhanced extractability of samples post-PEF is due to the dielectric disruption of cell membranes as a result of exposure to an electric current. This technology has already been explored as a non-thermal and, therefore, cheaper and more sustainable method of food preservation and microbial inactivation technique [77,78]. Outside of food preservation and antimicrobial applications, PEF has also been explored as a means of increasing juice yield from alfalfa leaves [79] and also as a pre-treatment to increase the extractability of proteins and phenolic compounds from light and dark BSG extracts [80]. This study by Kumari et al. (2019) showed that PEF assisted extraction significantly increased the level of free amino acids in the light BSG extract, and all essential amino acids were present in both extracts, with tryptophan being the only exception [80]. PEF treatment has also been explored in combination with ultrasound treatment as a method of optimising the extraction of phenolic compounds and proteins from agro-industrial by products [77] and olive kernels [81].

The more well-known methods of protein extraction, namely alkaline extraction and enzymatic treatment, are very suitable for protein extraction from BSG and are supported by large amounts of literature documenting their success. However, there are potential downsides and room for optimization with regards to the amount of solvent and/or enzyme required as well as heat treatments that could negatively impact protein quality. This leaves much room for optimization with a focus on cost, quality and sustainability. The more novel methods have little to no supporting literature with regards to their use for BSG but could benefit from being a 'cleaner' and more economical method of protein extraction, as there is no need for reagents, as well as potentially yielding protein of a higher quality due to a lack of thermal treatment. However, it is clear that the best course of action for protein extraction from BSG is a combination of the methods discussed above, as it has been shown in BSG and other food systems that combining methods allows for optimization of protein yield [68,70,72,77,81].

6. Applications

6.1. Animal Nutrition

Due to its high protein and fibre content, BSG is most often used as a component of animal feed as either a wet or dry feed. In combination with cheap and widely available nitrogen sources (such as urea), BSG can provide all essential amino acids [6]. The introduction of BSG into the diet of milking cattle has been shown to increase milk yield, total milk solids and milk fat when compared to an animal on a control diet. Protein levels and lactose content were not significantly affected by the change in diet [6,82–84]. While most commonly used in cattle, the use of BSG in the feed of other animals has also been explored. In a study by Mukasafari et al. (2017) BSG was successfully used to substitute up to 50% of sow and weaner meal without any negative effects on the quality of the pigs [85]. A study by Yaakugh and Tegbe (1994) investigated the replacement of maize in the diet of pigs with dried brewers' grains. It was found that, with a replacement level of up to 45%, the daily weight gain was significantly depressed, but the final carcass weight was not affected [86]. A study by Oh et al. (1991) investigated the incorporation of 15% treated BSG into the diets of poultry. The treatment consisted of partially hydrolysing the grains by cultivating the fungus *Trichoderma reesei* on the grains to alter the amino acid profile and release soluble sugars. Incorporation of these grains at 8 and 12% into the diets of broiler chicks resulted in a significant improvement in their growth and feed conversion ratio in the first 4 weeks. No further improvements were seen after 6 and 8 weeks of growth [87].

The use of BSG has been investigated as a source of protein in aquafeeds as a sustainable alternative for fishmeal and oil to reduce reliance on marine resources. Partial replacement of fishmeal with BSG (20–30%) in the feeding of rainbow trout and gilthead seabream showed similar results in digestion efficiency when compared to the control, where fishmeal was used as the main protein source. The BSG replacement showed good protein, amino acid and lipid digestibility, making BSG a suitable alternative for fishmeal, increasing the sustainability of both industries [3]. Similarly, San Martin et al. (2020) explored the use of an enzyme hydrolysis process in a bid to increase the digestibility of BSG proteins in aquaculture feeds. While further studies are required into the benefits of hydrolysis, both the hydrolysed and non-hydrolysed BSG proteins have shown good digestibility and would be a suitable fishmeal alternative, improving economic and environmental sustainability [2].

Due to the very high water content of BSG at the point of production (77–81%) [6], the transportation and storage of BSG for use as animal feed presents a challenge in minimizing microbial growth that may cause illness in animals, as well as general material degradation. There are guidelines regarding the preservation of BSG, suggesting the use of preservatives such as benzoate, propionate and sorbate to extend stability [6,10]. However, these measures only work to extend shelf-life for 4–5 days, so a more effective preservation technique is required for longer storage times. Drying is the most commonly employed methods of stabilizing wet BSG by reducing microbial growth. A study by Bartolome et al. (2002), compared three preservation methods: oven drying, freeze drying and freezing. Freezing was deemed to be unsuitable due to the potential for alterations in arabinose content and lack of suitability for large volumes. Oven drying and freeze drying were found to be equally effective in terms of reducing volume and preventing changes to the composition. However, from an economical and cost standpoint, oven drying was determined to be the most effective methods for removing moisture from BSG and stabilizing the product [88].

6.2. Bio-Degradable Film

A study by Lee et al. (2015) describes the use of brewers' spent grain protein (BGP) in the production of bio-degradable composite films. The brewers' spent grain protein was extracted by alkaline extraction, as described by Celus et al. (2007). It was determined that the addition of chitosan improved the physical and mechanical properties, including elasticity and tensile strength. The optimum levels of BGP and chitosan were determined to be 3 and 2%, respectively. In addition, the chitosan contributed to enhanced antimicrobial and antioxidant properties of the films, inhibiting the growth of *Staphylococcus aureus*, *Listeria monocytogenes*, *Escherichia coli* and *Salmonella typhimurium* [89]. The potential use of brewers' spent grain protein in biodegradable packaging materials is a very exciting potential use of the by-product.

Barley protein was also used to produce barley protein–gelatine composite films in a study by Song et al. (2012). The physical properties of this film were investigated, and it was found that increasing levels of barley bran protein caused tensile strength and elongation at break value to decrease. However, increasing the proportion of gelatine increased tensile strength but still reduced elongation value. It was determined that the optimal composition for film production was 3 g barley bran protein, 3 g gelatine and 100 mL sorbitol in 100 mL of film-forming solution. Grapefruit seed extract was also incorporated into the film as an antimicrobial agent and proved to be successful in reducing the growth of pathogenic bacteria when used in salmon packaging [90].

6.3. Food Applications

The implementation of brewers' spent yeast in bakery products has been widely studied. While these applications are not always protein-focused, in many cases, the protein fraction of BSG works synergistically with other compounds to improve nutritional and/or technological functionality. BSG addition to foodstuffs can improve protein content significantly but can also drastically increase levels of dietary fibre, which is very desirable for human health [48,91,92].

6.3.1. Biscuits

The use of protein isolated from brewers' spent grain has the potential for use in bakery products to improve nutritional and functional properties. A study by Zong et al. [93] implemented spent grain protein in cookies and studied the effects of the addition on the sensory and textural characteristics of the product. This addition improved the overall flavour and nutritional value of the cookies [77].

6.3.2. Bread

The use of BSG and fermented BSG in the fortification of bread has been explored by Waters et al. At 10% addition, dough displayed improved handling characteristics. As well as this, softness and staling were improved in both cases. While sweetness was decreased and acidulous flavour increased, both bread types were acceptable at a 10% addition level [48]. This acceptability, as well as the increase in nutritional value with regards to protein, dietary fibre and minerals, makes BSG a very interesting raw material for food product fortification.

The effect of BSG supplementation on bread doughs was also explored. In a study by Ktenioudaki et al. (2013), BSG and apple pomace were added to wheat dough, and their effect on the doughs physiochemical properties studied. The BSG was found to be high in protein (20.8%) and high in dietary fibre (60.5%). The rheological and pasting properties of the dough were greatly altered, mainly thought to be due to the high fibre content. Increasing by-product addition significantly reduced peak viscosity, holding strength, breakdown, final viscosity and setback values, as well as strain hardening index. Meanwhile, biaxial extension viscosity was higher for the supplemented dough and the storage modulus G'' was increased. These changes all indicate significant structural differences between the un-supplemented and supplemented doughs [94].

6.3.3. Snacks

Dried and milled BSG has been used as a means to increase the protein content of extruded snacks [95]. BSG was incorporated at 10–30% levels in combination with wheat flour, corn starch and other ingredients and extruded using a twin-screw extruder. This addition significantly increased crude protein content, as well as increasing the phytic acid level and bulk density. Meanwhile, sectional expansion and cell area were reduced. A follow-up study examining the effect of altering water feed rates using a combination of BSG and different flours was performed. The total dietary fibre of the wheat flour and BSG (WBSG) and cornflour and BSG (CBSG) mixtures were found to increase significantly. Generally, TDF increased with increasing water feed level for WBSG sample, while the opposite was observed with the CBSG samples [96].

A similar study by Ainsworth et al. (2007) also looked into the effects of BSG addition in the formulation for a chickpea-based extruded snack. It was determined that BSG addition (0–30%) resulted in decreased expansion, which is in agreement with other studies. However, increased screw speed had the opposite effect. The BSG product also displayed increased phytic acid, resistant starch and protein in vitro digestibility [97]. A study by Ktenioudaki, Crofton, et al. (2013) investigated the use of BSG in a crispy snack product as a means to increase fibre content. It was found that a 10% BSG addition almost doubled the fibre content of the product without compromising product acceptability [92]. While often used to increase protein levels in food products, BSG can also enhance other nutritional characteristics such as fibre content.

A study by Singh et al. investigated the potential use of novel drying methods on brewers' spent grain for use as a plant protein source in baked chips. Vacuum microwave drying (VMD) was proven to be an efficient method for drying BSG, as it reduced drying time, showed a high drying efficacy and a high overall acceptability in the baked snack when compared to those prepared using oven-dried or freeze-dried BSG. The applications of VMD technology are interesting from a sustainability standpoint as well as a nutritional standpoint, as the lack of a high temperature, due to the lowering of the boiling point by the vacuum, results in a higher-quality protein ingredient [98].

6.3.4. Beverages

The research surrounding the inclusion of BSG-derived ingredients in beverage formulation is limited. A study by McCarthy et al. (2013) investigated the use of phenolic compounds extracted from BSG as antioxidants in fruit beverages and determined that BSG extract addition resulted in significantly increased antioxidant activity, as measured by the ferric reducing antioxidant power (FRAP) assay [5]. While other studies incorporating this concept are rare, several patents are available (Table 2) regarding the extraction and use of BSG-derived ingredients in beverage applications.

Table 2. Patents regarding the use of BSG ingredients in food and beverage products.

Patent Number	Title	Area of Usage	Summary
WO/2018/033521	A process for preparing a beverage or beverage component, beverage or beverage component prepared by such process and use of brewers' spent grains for preparing such beverage or beverage component.	Food/beverage ingredient	Process of preparing a beverage or beverage component obtained by the fermentation of brewers' spent grain and a process of preparing such beverage and other foodstuffs.
WO/2018/033522	A process for preparing a beverage or beverage component from brewers' spent grains.	Food/beverage ingredient	Process of preparing a beverage or beverage component obtained by the enzymatic saccharification and fermentation of brewers' spent grain and a process of preparing such beverage, as well as other foodstuffs.
WO/2019/034567	A process for microbial stabilization of brewers' spent grain, microbiologically stabilized brewers' spent grain and use thereof.	Food/beverage ingredient	Process for treating brewers' spent grains (BSG) obtained from the brewing process such that the growth of microbes in said grains and subsequent production of microbial toxins are kept below specified levels.
WO/2019/158755	A process for recovering proteinaceous and/or fibrous material from brewers' spent grains and use thereof.	Food/beverage ingredient	Process of extracting or purifying proteinaceous material and/or fibraceous material from brewers' spent grain, as well as the use of these materials.

6.4. *Barley Protein Hydrolysates*

Due to a lack of solubility of barley proteins, functionality in food applications is limited. To improve ingredient functionality, protein hydrolysis can be implemented.

Celus et al. (2007) used enzymatic hydrolysis as a means to potentially improve the solubility, colour, emulsification and foaming properties, as well as cause a change in molecular weight distribution and hydrophobicity of BSG proteins. Differing degrees of hydrolysis were obtained by subjecting the protein concentrate to differing concentrations of three enzymes (Alcalase, Pepsin and Flavourzyme) for varying amounts of time. Generally, hydrolysis of the BSG protein improved emulsion and foaming. This somewhat agrees with a separate study by Yalçın et al. (2008), where barley protein hydrolysates were determined to have slightly improved foaming characteristics when compared to barley protein isolates, but the difference was slight [99]. However, those hydrolysates prepared with Alcalase and Pepsin showed a decrease in these characteristics with an increasing degree of hydrolysis. Characterisation of these hydrolysates showed that a relatively high molecular weight and a high surface hydrophobicity are desirable for enhanced physiochemical properties [64].

Enzymatic hydrolysis of barley proteins can greatly improve antioxidant capabilities and metal-binding activity [100–102]. A study by Chanput et al. (2009) examined the antioxidant properties of partially purified proteins from a variety of sources, including barley hordeins. Hydrolysates of these proteins were prepared using an enzymatic treatment of pepsin, followed by trypsin. Antioxidant and reducing properties were investigated, and the partially purified C-hordein displayed a high reducing capacity when compared to the B- and D-hordeins. It was generally found that for all protein hydrolysates, antioxidative and reducing capacities were greatly increased after enzymatic digestion with pepsin and trypsin [101]. A recent study by Ikram et al. (2020) followed on from this idea, investigating the effects of pre-treatments such as ultrasonic and heat treatments on the enzymatic hydrolysis of BSG proteins by Alcalase and the extent to which these pre-treatments, in combination with hydrolysis time, altered the antioxidant activity of the hydrolysates. The treatments selected were an ultrasonic treatment of 40 or 50 kHz and a heating treatment of 50 and 100 °C, while the pre-treatment times were 15, 30 and 60 min. The ultrasonic treatment at 40 and 50 kHz was shown to significantly increase oxygen radical absorption capacity values of the hydrolysates, while the heat treatment at 100 °C greatly increased the ferric reducing antioxidant power (FRAP) assay values [100]. These results indicate that hydrolysed BSG proteins exhibit a higher antioxidative power than untreated material and that these antioxidative capabilities can be further increased with ultrasonic and heat pre-treatments. Therefore, these hydrolysates could prove to be useful food ingredients due to their antioxidant and reducing potential.

Barley hordein hydrolysates also show potential as dietary supplements to enhance mineral bioavailability and solubility [102]. Through enzymatic digestion with Flavourzyme, Alcalase, trypsin and pepsin, barley proteins become hydrolysates with a strong metal ion binding capacity and can significantly increase the solubility of Fe^{2+} Fe^{2+}, Ca^{2+}, Cu^{2+} and Zn^{2+} [102]. Therefore, BSG proteins could potentially have a useful application in dietary supplements to enhance mineral bioavailability and solubility.

7. Conclusions

The increasing interest in sustainability and economising product side streams has led to an increased emphasis on the reuse and valorisation of brewing by-products. As brewers' spent grain is the most abundant by-product from this industry, the valorisation and utilization of spent grain protein is of great interest, particularly in terms of sustainability. Finding ways to upcycle this cheap and readily available product and apply it in a variety of different settings is a research area that is rapidly gaining traction. BSG is currently extremely underutilised and is mainly used in animal feed formulations due to its low cost and high nutritional value. High levels of essential amino acids in the proteins could be useful in nutritional and functional food applications for human consumption.

To date, BSG has been successfully applied in bakery products such as bread, biscuits and snack-type products, and BSG protein hydrolysates have been found to have increased functionality, including enhanced solubility, foaming and emulsification properties. These hydrolysates have also shown potential to enhance nutrient bioavailability as well as an increased antioxidative and reducing capability. All in all, BSG protein and its hydrolysates have significant valorisation potential, especially with regards to applications in the food industry.

Author Contributions: A.J.: writing, investigation; E.Z.: review/editing, funding acquisition; A.W.S.: project administration. E.K.A.: conceptualization, supervision, review/editing. All authors have read and agreed to the published version of the manuscript.

Funding: This project has received funding from the European Union's Horizon 2020 research and innovation programme under grant agreement No. 818368. This manuscript reflects only the authors' views and the European Commission is not responsible for any use that may be made of the information it contains.

Data Availability Statement: Not applicable.

Conflicts of Interest: The authors declare no conflict of interest.

References

1. Mussatto, S.I. Brewer' s Spent Grain: A Valuable Feedstock for Industrial Applications. *J. Sci. Food Agric.* **2014**, *94*, 1264–1275. [CrossRef]
2. San Martin, D.; Orive, M.; Iñarra, B.; Castelo, J.; Estévez, A.; Nazzaro, J.; Iloro, I.; Elortza, F.; Zufía, J. Brewers' Spent Yeast and Grain Protein Hydrolysates as Second-Generation Feedstuff for Aquaculture Feed. *Waste Biomass Valorization* **2020**, *11*, 5307–5320. [CrossRef]
3. Nazzaro, J.; Martin, D.S.; Perez-Vendrell, A.M.; Padrell, L.; Iñarra, B.; Orive, M.; Estévez, A. Apparent Digestibility Coefficients of Brewer's by-Products Used in Feeds for Rainbow Trout (Oncorhynchus Mykiss) and Gilthead Seabream (Sparus Aurata). *Aquaculture* **2021**, *530*, 735796. [CrossRef]
4. Mccarthy, A.L.; Callaghan, Y.C.O.; Piggott, C.O.; Fitzgerald, R.J.; Brien, N.M.O. Postgraduate Symposium Brewers ' Spent Grain; Bioactivity of Phenolic Component, Its Role in Animal Nutrition and Potential for Incorporation in Functional Foods: A Review. *Proc. Nutr. Soc.* **2013**, *72*, 117–125. [CrossRef] [PubMed]
5. McCarthy, A.L.; O'Callaghan, Y.C.; Neugart, S.; Piggott, C.O.; Connolly, A.; Jansen, M.A.K.; Krumbein, A.; Schreiner, M.; FitzGerald, R.J.; O'Brien, N.M. The Hydroxycinnamic Acid Content of Barley and Brewers' Spent Grain (BSG) and the Potential to Incorporate Phenolic Extracts of BSG as Antioxidants into Fruit Beverages. *Food Chem.* **2013**, *141*, 2567–2574. [CrossRef] [PubMed]
6. Mussatto, S.I.; Dragone, G.; Roberto, I.C. Brewers ' Spent Grain: Generation, Characteristics and Potential Applications. *J. Cereal Sci.* **2006**, *43*, 1–14. [CrossRef]
7. Fox, G.P. *Chemical Composition in Barley Grains and Malt Quality*; Springer: Berlin/Heidelberg, Germany, 2010; ISBN 9783642012792.
8. Jacobsen, J.V.; Knox, R.B.; Pyliotis, N.A. The Structure and Composition of Aleurone Grains in the Barley Aleurone Layer. *Planta* **1971**, *101*, 189–209. [CrossRef]
9. Arendt, E.K.; Zannini, E. *Cereal Grains for the Food and Beverage Industries*; Elsevier: Amsterdam, The Netherlands, 2013; ISBN 9780857094131.
10. Lynch, K.M.; Steffen, E.J.; Arendt, E.K. Brewers' Spent Grain: A Review with an Emphasis on Food and Health. *J. Inst. Brew.* **2016**. [CrossRef]
11. Agu, R.C.; Bringhurst, T.A.; Brosnan, J.M.; Pearson, S. Potential of Hull-Less Barley Malt for Use in Malt and Grain Whisky Production. *J. Inst. Brew.* **2009**, *115*, 128–133. [CrossRef]
12. Gupta, M.; Abu-Ghannam, N.; Gallaghar, E. Barley for Brewing: Characteristic Changes during Malting, Brewing and Applications of Its by-Products. *Compr. Rev. Food Sci. Food Saf.* **2010**, *9*, 318–328. [CrossRef]
13. Vieira, E.; Rocha, M.A.M.; Coelho, E.; Pinho, O.; Saraiva, J.A.; Ferreira, I.M.P.L.V.O.; Coimbra, M.A. Valuation of Brewer's Spent Grain Using a Fully Recyclable Integrated Process for Extraction of Proteins and Arabinoxylans. *Ind. Crops Prod.* **2014**, *52*, 136–143. [CrossRef]
14. Qi, J.C.; Zhang, G.P.; Zhou, M.X. Protein and Hordein Content in Barley Seeds as Affected by Nitrogen Level and Their Relationship to Beta-Amylase Activity. *J. Cereal Sci.* **2006**, *43*, 102–107. [CrossRef]
15. Howard, K.A.; Gayler, K.R.; Eagles, H.A.; Halloran, G.M. The Relationship between D Hordein and Malting Quality in Barley. *J. Cereal Sci.* **1996**, *24*, 47–53. [CrossRef]
16. Eagles, H.A.; Bedggood, A.G.; Panozzo, J.F.; Martin, P.J. Cultivar and Environmental Effects on Malting Quality in Barley. *Aust. J. Agric. Res.* **1995**, *46*, 831–844. [CrossRef]
17. Weston, D.T.; Horsley, R.D.; Schwarz, P.B.; Goos, R.J. Nitrogen and Planting Date Effects on Low-Protein Spring Barley. *Agron. J.* **1993**, *85*, 1170–1174. [CrossRef]
18. Zhang, G.; Chen, J.; Wang, J.; Ding, S. Cultivar and Environmental Effects on $(1 \rightarrow 3, 1 \rightarrow 4)$ β-D-Glucan and Protein Content in Malting Barley. *J. Cereal Sci.* **2001**, *34*, 295–301. [CrossRef]
19. Cai, S.; Yu, G.; Chen, X.; Huang, Y.; Jiang, X.; Zhang, G.; Jin, X. Grain Protein Content Variation and Its Association Analysis in Barley. *BMC Plant Biol.* **2013**, *13*. [CrossRef]
20. Mahalingam, R. Phenotypic, Physiological and Malt Quality Analyses of US Barley Varieties Subjected to Short Periods of Heat and Drought Stress. *J. Cereal Sci.* **2017**, *76*, 199–205. [CrossRef]
21. Shewry, P.R. Improving the Protein Content and Composition of Cereal Grain. *J. Cereal Sci.* **2007**, *46*, 239–250. [CrossRef]
22. Baxter, E.D. Hordein in Barley and Malt—A Review. *J. Inst. Brew.* **1981**, *87*, 173–176. [CrossRef]

23. Shewry, P.R.; Parmar, S.; Field, J.M. Two-dimensional Electrophoresis of Cereal Prolamins: Applications to Biochemical and Genetic Analyses. *Electrophoresis* **1988**, *9*, 727–737. [CrossRef]
24. Shewry, P.R.; Kreis, M.; Parmar, S.; Lew, E.J.L.; Kasarda, D.D. Identification of γ-Type Hordeins in Barley. *FEBS Lett.* **1985**, *190*, 61–64. [CrossRef]
25. Zhao, J.; Tian, Z.; Chen, L. Effects of Deamidation on Aggregation and Emulsifying Properties of Barley Glutelin. *Food Chem.* **2011**, *128*, 1029–1036. [CrossRef]
26. Wang, C.; Tian, Z.; Chen, L.; Temelli, F.; Liu, H.; Wang, Y. Functionality of Barley Proteins Extracted and Fractionated by Alkaline and Alcohol Methods. *Cereal Chem.* **2010**, *87*, 597–606. [CrossRef]
27. Klose, C.; Thiele, F.; Arendt, E.K. Changes in the Protein Profile of Oats and Barley during Brewing and Fermentation. *J. Am. Soc. Brew. Chem.* **2010**, *68*, 119–124. [CrossRef]
28. Niu, C.; Han, Y.; Wang, J.; Zheng, F.; Liu, C.; Li, Y.; Li, Q. Malt Derived Proteins: Effect of Protein Z on Beer Foam Stability. *Food Biosci.* **2018**, 21–27. [CrossRef]
29. Ayashi, K.A.H.; To, K.A.I.; Ato, K.A.S.; Akeda, K.A.T. Novel Prediction Method of Beer Foam Stability Using Protein Z, Barley Dimeric r -Amylase Inhibitor-1 (BDAI-1) and Yeast Thioredoxin. *J. Agric. Food Chem.* **2008**, *1*, 8664–8671.
30. Kapp, G.R.; Bamforth, C.W. The Foaming Properties of Proteins Isolated from Barley. *J. Sci. Food Agric.* **2002**, *1281*, 1276–1281. [CrossRef]
31. Boba, J. Monitoring of Malting Process by Characterization of Glycation of Barley Protein Z. *Eur. Food Res. Technol.* **2010**, 665–673. [CrossRef]
32. Celus, I.; Brijs, K.; Delcour, J.A. The Effects of Malting and Mashing on Barley Protein Extractability. *J. Cereal Sci.* **2006**, *44*, 203–211. [CrossRef]
33. Osman, A.M.; Coverdale, S.M.; Cole, N.; Hamilton, S.E.; de Jersey, J. Characterisation and Assessment of the Role of Barley Malt Endoproteases During Malting and Mashing1. *J. Inst. Brew.* **2002**, *108*, 62–67. [CrossRef]
34. Slack, P.T.; Baxter, E.D.; Wainwright, T. Inhibition By Hordein of Starch Degradation. *J. Inst. Brew.* **1979**, *85*, 112–114. [CrossRef]
35. Yu, W.; Zou, W.; Dhital, S.; Wu, P.; Gidley, M.J.; Fox, G.P.; Gilbert, R.G. The Adsorption of α-Amylase on Barley Proteins Affects the in Vitro Digestion of Starch in Barley Flour. *Food Chem.* **2018**, *241*, 493–501. [CrossRef] [PubMed]
36. Jones, B.L.; Marinac, L. The Effect of Mashing on Malt Endoproteolytic Activities. *J. Agric. Food Chem.* **2002**, *50*, 858–864. [CrossRef]
37. Steiner, E.; Gastl, M.; Becker, T. Protein Changes during Malting and Brewing with Focus on Haze and Foam Formation: A Review. *Eur. Food Res. Technol.* **2011**, *232*, 191–204. [CrossRef]
38. Moonen, J.H.E.; Graveland, A.; Muts, G.C. The molecular structure of gelprotein from barley, its behaviour in wort—Filtration and analysis. *J. Inst. Brew.* **1987**, *93*, 125–130. [CrossRef]
39. Connolly, A.; Piggott, C.O.; Fitzgerald, R.J. Characterisation of Protein-Rich Isolates and Antioxidative Phenolic Extracts from Pale and Black Brewers' Spent Grain. *Int. J. Food Sci. Technol.* **2013**, *48*, 1670–1681. [CrossRef]
40. Skerritt, J.H.; Janes, P.W. Disulphide-Bonded 'Gel Protein' Aggregates in Barley: Quality-Related Differences in Composition and Reductive Dissociation. *J. Cereal Sci.* **1992**, *16*, 219–235. [CrossRef]
41. Curioni, A.; Pressi, G.; Furegon, L.; Peruffo, A.D.B.; Agrarie, B.; Gradenigo, V. Major Proteins of Beer and Their Precursors in Barley: Electrophoretic and Immunological Studies. *J. Agric. Food Chem.* **1995**, 2620–2626. [CrossRef]
42. Kanauchi, O.; Araki, Y.; Andoh, A.; Mitsuyama, K. Prebiotic Treatment of Experimental Colitis with Germinated Barley Foodstuff: A Comparison with Probiotic or Antibiotic Treatment. *Int. J. Mol. Med.* **2002**, *1*, 65–70.
43. Xiros, C.; Topakas, E.; Katapodis, P.; Christakopoulos, P. Hydrolysis and Fermentation of Brewer' s Spent Grain by Neurospora Crassa. *Bioresour. Technol.* **2008**, *99*, 5427–5435. [CrossRef]
44. Jay, A.J.; Parker, M.L.; Faulks, R.; Husband, F.; Wilde, P.; Smith, A.C.; Faulds, C.B.; Waldron, K.W. A Systematic Micro-Dissection of Brewers ' Spent Grain. *J. Cereal Sci.* **2008**, *47*, 357–364. [CrossRef]
45. Robertson, J.A.; Anson, K.J.A.I.; Treimo, J.; Faulds, C.B.; Brocklehurst, T.F.; Eijsink, V.G.H.; Waldron, K.W. LWT—Food Science and Technology Profiling Brewers ' Spent Grain for Composition and Microbial Ecology at the Site of Production. *LWT Food Sci. Technol.* **2010**, *43*, 890–896. [CrossRef]
46. Roberto, C. Mussatto Chemical Characterization and Liberation of Pentose Sugars from Brewer' s Spent Grain. *J. Chem. Technol. Biotechnol. Int. Res. Process. Environ. Clean Technol.* **2006**, *274*, 268–274. [CrossRef]
47. Santos, M.; Jimenez, J.; Bartolome, B.; Gomez-Cordoves, C.; Del Nozal, M. Variability of Brewer' s Spent Grain within a Brewery. *Food Chem.* **2003**, *80*, 17–21. [CrossRef]
48. Waters, D.M.; Jacob, F.; Titze, J.; Arendt, E.K.; Zannini, E. Fibre, Protein and Mineral Fortification of Wheat Bread through Milled and Fermented Brewer ' s Spent Grain Enrichment. *Eur. Food Res.Technol.* **2012**, 767–778. [CrossRef]
49. Meneses, N.G.T.; Martins, S.; Teixeira, J.A.; Mussatto, S.I. Influence of Extraction Solvents on the Recovery of Antioxidant Phenolic Compounds from Brewer' s Spent Grains. *Sep. Purif. Technol.* **2013**, *108*, 152–158. [CrossRef]
50. Bogdan, P.; Kordialik-bogacka, E. Trends in Food Science & Technology Alternatives to Malt in Brewing. *Trends Food Sci. Technol.* **2017**, *65*, 1–9. [CrossRef]
51. Yorke, J.; Cook, D.; Ford, R. Brewing with Unmalted Cereal Adjuncts: Sensory and Analytical Impacts on Beer Quality. *Beverages* **2021**, *7*, 4. [CrossRef]

52. Lloyd, B.W.J.W. Amount of Starch or Sugar from the Agricultural Source. Hydrate Source38 Other than Malted Barley Which Contributes Case of Extraction and the Amount of Nitrogenous Materials Using Isothermal Conversion. *J. Inst. Brew* **1986**, *92*, 336–345. [CrossRef]
53. Glatthar, J.; Heinisch, J.; Senn, T. Unmalted Triticale Cultivars as Brewing Adjuncts: Effects of Enzyme Activities and Composition on Beer Wort Quality. *J. Sci. Food Agric.* **2005**, *654*, 647–654. [CrossRef]
54. Agu, R.C. A Comparison of Maize, Sorghum and Barley as Brewing Adjuncts. *J. Inst. Brew.* **2002**, *108*, 19–22. [CrossRef]
55. Schnitzenbaumer, B.; Kerpes, R.; Titze, J.; Jacob, F.; Arendt, E. Impact of Various Levels of Unmalted Oats (*Avena sativa* L.) on the Quality and Processability of Mashes. Worts, and Beers. *J. Am. Soc. Brew. Chem.* **2012**, *70*, 142–149. [CrossRef]
56. Marconi, O.; Sileoni, V.; Marconi, O.; Sileoni, V.; Ceccaroni, D.; Perretti, G. The Use of Rice in Brewing. *Adv. Int. Rice Res.* **2017**, 49–66. [CrossRef]
57. Zhou, Z.; Robards, K.; Helliwell, S.; Blanchard, C. Review Composition and Functional Properties of Rice. *Int. J. food Sci. Technol.* **2002**, *37*, 849–868. [CrossRef]
58. Taylor, J.R.N.; Dlamini, B.C.; Kruger, J. 125 Th Anniversary Review: The Science of the Tropical Cereals Sorghum, Maize and Rice in Relation to Lager Beer Brewing. *J. Inst. Brew.* **2013**, 1–14. [CrossRef]
59. Taylor, P.; Bechtel, D.B. Properties of Whole and Undigested Fraction of Protein Bodies of Milled Rice Preparations Had Higher Fat Content than Whole PB Preparation. Endosperm Protein of Rice Exists Mainly as That the Protein Fraction Rendered Undigested. *Agric. Biol. Chem.* **2015**, *42*, 2015–2023. [CrossRef]
60. Watson, S. Description, development, structure, and composition of the corn kernel. In *Corn: Chemistry and Technology*; American Association of Cereal Chemists: Saint Paul, MN, USA, 2003; pp. 69–106. ISBN 1891127330.
61. Giese, H.; Hejgaard, J. Synthesis of Salt-Soluble Proteins in Barley. Pulse-Labeling Study of Grain Filling in Liquid-Cultured Detached Spikes. *Planta* **1984**, *161*, 172–177. [CrossRef]
62. Shewry, P.R.; Ellis, J.R.S.; Pratt, H.M.; Miflin, B.J. A Comparison of Methods for the Extraction and Separation of Hordein Fractions from 29 Barley Varieties. *J. Sci. Food Agric.* **1978**, *29*, 433–441. [CrossRef]
63. Cavonius, L.R.; Albers, E.; Undeland, I. PH-Shift Processing of Nannochloropsis Oculata Microalgal Biomass to Obtain a Protein-Enriched Food or Feed Ingredient. *ALGAL* **2015**, *11*, 95–102. [CrossRef]
64. Celus, I.; Brijs, K.; Delcour, J.A. Enzymatic Hydrolysis of Brewers' Spent Grain Proteins and Technofunctional Properties of the Resulting Hydrolysates. *J. Agric. Food Chem.* **2007**, *55*, 8703–8710. [CrossRef]
65. Bals, B.; Balan, V.; Dale, B. Bioresource Technology Integrating Alkaline Extraction of Proteins with Enzymatic Hydrolysis of Cellulose from Wet Distiller' s Grains and Solubles. *Bioresour. Technol.* **2009**, *100*, 5876–5883. [CrossRef]
66. Diptee, R.; Smith, J.P.; Alli, I.; Khanizadeh, S. Application of response surface methodology in protein extraction studies from brewer's spent grain. *J. Food Process. Preserv.* **1989**, *13*, 457–474. [CrossRef]
67. Templin, T.L.; Johnston, D.B.; Singh, V.; Tumbleson, M.E.; Belyea, R.L.; Rausch, K.D. Membrane Separation of Solids from Corn Processing Streams. *Bioresour. Technol.* **2006**, *97*, 1536–1545. [CrossRef] [PubMed]
68. Tang, D.S.; Yin, G.M.; He, Y.Z.; Hu, S.Q.; Li, B.; Li, L.; Liang, H.L.; Borthakur, D. Recovery of Protein from Brewer's Spent Grain by Ultrafiltration. *Biochem. Eng. J.* **2009**, *48*, 1–5. [CrossRef]
69. Connolly, A.; Cermeno, M.; Crowley, D.; O'Callaghan, Y.; O'Brien, N.M.; Fitzgerald, R.J. Characterisation of the in Vitro Bioactive Properties of Alkaline and Enzyme Extracted Brewers' Spent Grain Protein Hydrolysates. *Food Res. Int.* **2019**, *121*, 524–532. [CrossRef]
70. Qin, F.; Johansen, A.Z.; Mussatto, S.I. Evaluation of Different Pretreatment Strategies for Protein Extraction from Brewer's Spent Grains. *Ind. Crops Prod.* **2018**, *125*, 443–453. [CrossRef]
71. Niemi, P.; Martins, D.; Buchert, J.; Faulds, C.B. Pre-Hydrolysis with Carbohydrases Facilitates the Release of Protein from Brewer's Spent Grain. *Bioresour. Technol.* **2013**, *136*, 529–534. [CrossRef]
72. Rommi, K.; Niemi, P.; Kemppainen, K.; Kruus, K. Impact of Thermochemical Pre-Treatment and Carbohydrate and Protein Hydrolyzing Enzyme Treatment on Fractionation of Protein and Lignin from Brewer's Spent Grain. *J. Cereal Sci.* **2018**, *79*, 168–173. [CrossRef]
73. Treimo, J.; Aspmo, S.I.; Eijsink, V.G.H.; Horn, S.J. Enzymatic Solubilization of Proteins in Brewer's Spent Grain. *J. Agric. Food Chem.* **2008**, *56*, 5359–5365. [CrossRef] [PubMed]
74. Tang, D.S.; Tian, Y.J.; He, Y.Z.; Li, L.; Hu, S.Q.; Li, B. Optimisation of Ultrasonic-Assisted Protein Extraction from Brewer's Spent Grain. *Czech J. Food Sci.* **2010**, *28*, 9–17. [CrossRef]
75. Li, W.; Yang, H.; Emilia, T.; Zhao, H. Modification of Structural and Functional Characteristics of Brewer's Spent Grain Protein by Ultrasound Assisted Extraction. *LWT* **2020**, *139*, 110582. [CrossRef]
76. Yu, D.; Sun, Y.; Wang, W.; Keefe, S.F.O.; Neilson, A.P.; Feng, H.; Wang, Z. Original Article Recovery of Protein Hydrolysates from Brewer ' s Spent Grain Using Enzyme and Ultrasonication. *Int. J. Food Sci. Technol.* **2019**, *55*, 357–368. [CrossRef]
77. Kumari, B.; Tiwari, B.K.; Hossain, M.B.; Brunton, N.P.; Rai, D.K. Recent Advances on Application of Ultrasound and Pulsed Electric Field Technologies in the Extraction of Bioactives from Agro-Industrial By-Products. *Food Bioprocess Technol.* **2017**. [CrossRef]
78. Arshad, R.N.; Abdul-Malek, Z.; Roobab, U.; Munir, M.A.; Naderipour, A.; Qureshi, M.I.; Bekhit, A.; Liu, Z.W.; Aadil,, R.M. Pulsed Electric Field: A Potential Alternative towards a Sustainable Food Processing. *Trends Food Sci. Technol.* **2021**, *111*, 43–54. [CrossRef]

79. Gachovska, T.K.; Ngadi, M.O.; Raghavan, G.S.V. Pulsed Electric Field Assisted Juice Extraction from Alfalfa. *Can. Biosyst. Eng.* **2006**, *48*, 3.
80. Kumari, B.; Tiwari, B.K.; Walsh, D.; Griffin, T.P.; Islam, N.; Lyng, J.G.; Brunton, N.P.; Rai, D.K.; Biosciences, F.; Food, T.; et al. Impact of Pulsed Electric Field Pre-Treatment on Nutritional and Polyphenolic Contents and Bioactivities of Light and Dark Brewer ' s Spent Grains. *Innov. Food Sci. Emerg. Technol.* **2019**, *54*, 200–210. [CrossRef]
81. Roselló-soto, E.; Barba, F.J.; Parniakov, O.; Galanakis, C.M. High Voltage Electrical Discharges, Pulsed Electric Field, and Ultrasound Assisted Extraction of Protein and Phenolic Compounds from Olive Kernel. *Food Bioprocess Technol.* **2014**, *8*, 885–894. [CrossRef]
82. Belibasakis, N.G.; Tsirgogianni, D. Effects of Wet Brewers Grains on Milk Yield, Milk Composition and Blood Components of Dairy Cows in Hot Weather. Anim. *Feed Sci. Technol.* **1996**, *57*, 175–181. [CrossRef]
83. Getu, K. Supplementary Value of Ensiled Brewers Spent Grain Used as Replacement to Cotton Seed Cake in the Concentrate Diet of Lactating Crossbred Dairy Cows. Trop. *Anim. Health Prod.* **2020**, *52*, 3675–3683. [CrossRef]
84. Ikram, S.; Huang, L.; Zhang, H.; Wang, J. Composition and Nutrient Value Proposition of Brewers Spent Grain. *J. Food Sci.* **2017**, *82*. [CrossRef]
85. Mukasafari, M.A.; Ambula, M.K.; Karege, C.; King, A.M. Effects of Substituting Sow and Weaner Meal with Brewers' Spent Grains on the Performance of Growing Pigs in Rwanda. *Trop. Anim. Health Prod.* **2017**, *50*, 393–398. [CrossRef]
86. Yaakugh, I.D.I.; Tegbe, B. Replacement Value of Brewers ' Dried Grain for Maize on Performance of Pigs. *J. Sci. Food* **1994**, 465–471. [CrossRef]
87. Oh, J.C.S.; Chngt, A.L.; Jesudason, R.B. Incorporation of Microbiologically Treated Spent Brewery Grains into Broiler Rations. *Lett. Appl. Microbiol.* **1991**, *13*, 150–153. [CrossRef]
88. Bartolome, B.; Santos, M.; Jime, J.J.; Nozal, M.J.; Go, C. Pentoses and Hydroxycinnamic Acids in Brewer' s Spent Grain. *J. Cereal Sci.* **2002**, *36*, 51–58. [CrossRef]
89. Lee, J.H.; Lee, J.H.; Yang, H.J.; Song, K. Bin Preparation and Characterization of Brewer's Spent Grain Protein-Chitosan Composite Films. *J. Food Sci. Technol.* **2015**, *52*, 7549–7555. [CrossRef]
90. Song, H.Y.; Shin, Y.J.; Song, K. Bin Preparation of a Barley Bran Protein—Gelatin Composite Film Containing Grapefruit Seed Extract and Its Application in Salmon Packaging. *J. Food Eng.* **2012**, *113*, 541–547. [CrossRef]
91. Stojceska, V.; Ainsworth, P. The Effect of Different Enzymes on the Quality of High-Fibre Enriched Brewer' s Spent Grain Breads. *Food Chem.* **2008**, *110*, 865–872. [CrossRef] [PubMed]
92. Ktenioudaki, A.; Crofton, E.; Scannell, A.G.M.; Hannon, J.A.; Kilcawley, K.N.; Gallagher, E. Sensory Properties and Aromatic Composition of Baked Snacks Containing Brewer's Spent Grain. *J. Cereal Sci.* **2013**, *57*, 384–390. [CrossRef]
93. Zong, X.Y.; Bian, M.H.; Li, L.; Qiang, L.; Luo, H. Study of Protein Cookies with Brewer's Grains. *J. Sichuan Univ. Sci. Eng.* **2012**, *4*, 14–16.
94. Ktenioudaki, A.; Shea, N.O.; Gallagher, E. Rheological Properties of Wheat Dough Supplemented with Functional By-Products of Food Processing: Brewer ' s Spent Grain and Apple Pomace. *J. Food Eng.* **2013**, *116*, 362–368. [CrossRef]
95. Stojceska, V.; Ainsworth, P.; Plunkett, A.; Ibanoğlu, S. The Recycling of Brewer's Processing by-Product into Ready-to-Eat Snacks Using Extrusion Technology. *J. Cereal Sci.* **2008**, *47*, 469–479. [CrossRef]
96. Stojceska, V.; Ainsworth, P.; Plunkett, A.; Ibanoğlu, Ş. The Effect of Extrusion Cooking Using Different Water Feed Rates on the Quality of Ready-to-Eat Snacks Made from Food by-Products. *Food Chem.* **2009**, *114*, 226–232. [CrossRef]
97. Ainsworth, P.; Ibanoğlu, Ş.; Plunkett, A.; Ibanoğlu, E.; Stojceska, V. Effect of Brewers Spent Grain Addition and Screw Speed on the Selected Physical and Nutritional Properties of an Extruded Snack. *J. Food Eng.* **2007**, *81*, 702–709. [CrossRef]
98. Singh, A.P.; Mandal, R.; Shojaei, M.; Singh, A.; Kowalczewski, P.L.; Ligaj, M.; Pawlicz, J.; Jarzebski, M. Novel Drying Methods for Sustainable Upcycling of Brewers' Spent Grains as a Plant Protein Source. *Sustainability* **2020**, *12*, 3660. [CrossRef]
99. Yalçın, E.; Çelik, S.; İbanoğlu, E. Foaming Properties of Barley Protein Isolates and Hydrolysates. *Eur. Food Res. Technol.* **2008**, *226*, 967–974. [CrossRef]
100. Ikram, S.; Zhang, H.; Ahmed, M.S.; Wang, J. Ultrasonic Pretreatment Improved the Antioxidant Potential of Enzymatic Protein Hydrolysates from Highland Barley Brewer's Spent Grain (BSG). *J. Food Sci.* **2020**, *85*, 1045–1059. [CrossRef] [PubMed]
101. Chanput, W.; Theerakulkait, C.; Nakai, S. Antioxidative Properties of Partially Purified Barley Hordein, Rice Bran Protein Fractions and Their Hydrolysates. *J. Cereal Sci.* **2009**, *49*, 422–428. [CrossRef]
102. Eckert, E.; Bamdad, F.; Chen, L. Metal Solubility Enhancing Peptides Derived from Barley Protein. *Food Chem.* **2014**, *159*, 498–506. [CrossRef]

MDPI
St. Alban-Anlage 66
4052 Basel
Switzerland
Tel. +41 61 683 77 34
Fax +41 61 302 89 18
www.mdpi.com

Foods Editorial Office
E-mail: foods@mdpi.com
www.mdpi.com/journal/foods

www.ingramcontent.com/pod-product-compliance
Lightning Source LLC
LaVergne TN
LVHW070652170726
843515LV00004B/388
* 9 7 8 3 0 3 6 5 4 6 3 1 5 *